W9-CHK-829

Evaluation Methods for Environmental Standards

Principal Author

William D. Rowe
Director, Institute for Risk Analysis
Professor, Center for Technology and Administration
The American University
Washington, D.C.

Editor

Frederick J. Hageman
Environmental Protection Agency
Washington, D.C.

Contributing Authors

Ruth H. Allen
Frederick J. Hageman
Annette T. MacIntyre
John K. Overbey

CRC Press, Inc.
Boca Raton, Florida

Library of Congress Cataloging in Publication Data

Rowe, William D., 1930-
    Risk assessment approaches in establishing
environmental health standards.

    Bibliography: p.
    Includes index.
    1. Environmental health--Standards. 2. Toxicity
testing. 3. Cadmium--Toxicology. 4. Phenols--
Toxicology. 5. Water--Pollution--Toxicology.
I. Title.
RA566.R68   1983   363.7'394   82-24341
ISBN 0-8493-5967-8

    Direct all inquiries to CRC Press, Inc., 2000 Corporate Blvd., N.W., Boca Raton, Florida, 33431.

© 1983 by CRC Press, Inc.
Second Printing, 1984

International Standard Book Number 0-8493-5967-8

Library of Congress Card Number 82-24341
Printed in the United States

PREFACE

The genesis of this book was a study conducted by the American University Institute for Risk Analysis (AURA) for the American Iron and Steel Institute (AISI). This study, addressing methods of setting levels of standard for water pollution, was phase I of an evaluation of risk assessment problems and methods for all media.

The study provided useful results, which, as outlined in the Introduction, we believe to be of major importance to those preparing and living with all types of standards. Because of this, we decided to reassemble the material in a form which might benefit and be readily used by the student of risk analysis, as well as by regulators and "regulatees".

The objective of the original study, as described in the ensuing chapters, was to evaluate the process of setting acceptable levels of risk in environmental standards as it is used in regulatory practice today and how it may be used in the future. We selected two toxic substances, cadmium and phenol, as test vehicles because:

- Both are chemically toxic at high dose levels and may be carcinogenic at low levels
- One is inorganic and the other organic
- Sufficient data exist to analyse risk estimates and gauge their limitations
- Both are in wide use and are not specific to a particular industry

Other substances could have been chosen, but our purpose was not to assess the risks of these chemicals per se, but to use them to evaluate the methods and problems involved in risk assessment. On this basis, a protocol was established to evaluate the risk assessment process. We trust that the application of ideas from this protocol (briefly outlined in the Introduction) and the principles discussed will assist the reader in his or her endeavors in this area.

The initial project, funded through a grant from the American Iron and Steel Institute, resulted in a rather complex technical report. William D. Rowe, Ph.D., the principal investigator on the project, is also the principal author of the technical report and the book in its present form. Frederick Hageman has been responsible for converting the initial report into book form as prsented here. Mr. Hageman also contributed to those portions of the study involving biota other than man and human exposure. Devra Davis, Ph.D. and Harvey Babich, Ph.D., were responsible for gathering the health related and environmental fate information on cadmium and phenol. Ruth Allen, Ph.D. used this information in conjunction with the principal investigator to develop the dose/effect relationships found in Part V. Annette T. MacIntyre was responsible for developing the information on cadmium producing industries and their control, as described in Part V, and John K. Overbey was responsible for phenol producing industries and their controls as well as development of the river flow and dilution models used in both Part III and Part IV.

We also wish to point out that the American Iron and Steel Institute provided helpful comments through the efforts of A. E. Moffitt, Jr., Bernard F. Schneider, Ph.D., D., Gary Hancock, and E. F. Young, Jr., in terms of editing to final form. It should be emphasized that the study results as reported here are solely those of American University Institute for Risk Analysis (AURA) and the authors.

William D. Rowe, Ph.D.
June 1982

INTRODUCTION

It is not enough that you should understand about applied science in order that your work may increase man's blessings. Concern for man himself and his fate must always form the chief interest of all technical endeavors, concern for the great unsolved problems of the organization of labor and the distribution of goods — in order that the creations of our mind shall be a blessing and not a curse to mankind. Never forget this in the midst of your diagrams and equations.

Albert Einstein
Address, California Institute of Technology
(1931)

This book explores and evaluates the many approaches to setting acceptable levels of risk in environmental regulatory standards. Its usefulness is not limited to the area of standard setting, however. Anyone concerned about, or involved in, the process of risk evaluation may find useful principles and techniques in these pages.

Disciplines required for risk analysis have given us the means for providing betteer perspectives on environmental risks, but have not alleviated the need for value judgments in establishing acceptable levels for regulatory standards. To evaluate quantitative approaches to and needed value judgments in standard setting, we investigated the effects of two substances discharged into water. We selected cadmium because it is both toxic and carcinogenic; it persists in the environment and it is inorganic. The other substance, phenol, was chosen because it is highly toxic in high concentration, there is suggestive evidence that it may be carcinogenic, it biodegrades readily in the environment, and it is organic.

I. OVERVIEW OF THE STUDY FORMAT

Using published data, we developed health impact estimates for both pollutants based on acute and chronic effects. Throughout, we have separated carcinogenic from non-carcinogenic chronic risks because of differences: in how the public perceives the risk from each, in regulatory policies, and in dose-effect relationships. The ranges of uncertainty of health impact were preserved in establishing these estimates.

To assist the reader, we have presented this material in a logical sequence which follows a typical process that might be involved in developing a standard. For example, to initiate the analysis, one must have some background knowledge of the discharging industries and must select models to estimate the amount of this discharge and the portion reaching man. The information is presented first. The investigation to obtain this data also provides us with data on the value of the technological products produced by these industries. These data are used to estimate the value of loss of production capability, if any, resulting from applications of standards.

Under certain conditions, it is conceivable that other sequences of discrete tasks might be desirable or necessary to the investigator. It is our intent to provide the reader with both a framework for understanding or critiquing methods of setting levels for standards and an overview of the standard setting process.

We next evaluated risk from all sources of cadmium and phenol by estimating discharge rates by various industries into the environment. This information was used in a river dilution model and a population exposure model, so that exposure from discharges in terms of dose could be ascertained. Finally, we made risk estimates from the resulting exposure levels, using previously developed health impact estimates in the form of dose-response relationships.

We then estimated the effectiveness and costs of alternative controls for each industry and evaluated both the value of the products of the industrial technology and the availability and suitability of substitutes. Uncertainties are retained throughout this part of the analysis as well.

Our objective is not to precisely assess the environmental risks of cadmium and phenol in water, but to provide a useful vehicle for analyzing approaches to setting both standards and levels of acceptable risk. In this respect, the estimates may or may not coincide with reality although we hope they are reasonably accurate. They should not be cited as risk estimates, merely as vehicles for the evaluation of methods of setting acceptable levels of risk in standards.

Using these vehicles, different methods of setting acceptable levels of risk for different types of standards are evaluated. We have adopted a number of different types of standards which seem reasonably applicable for control of effluents in water, and applied a range of alternative methods to establish acceptable levels of risk for each to determine relative impact on risk reduction. We then evaluated each of these methods in terms of the risk reduced, the direct and indirect costs, the value of technology affected, and the cost-effectiveness of risk reduction. In each case, we evaluated the sensitivity of the methods in light of uncertainties in the risk estimates and the required value judgments. Finally, we present an overview of the approaches and an evaluation of the methods of setting acceptable risk levels. Realizing our study limitations and the restriction on two substances in water, we were still able to draw a number of important conclusions, which we believe are generally applicable.

II. OVERVIEW OF STUDY CONCLUSIONS

Conclusions are

- The situation determines the method or methods to be used in establishing levels for standards. Conversely, there are no universally applicable methods. Moreover, the pursuit and attempted use of believed universally applicable methods is wasteful, cost-ineffective, and often counter-productive.
- In the absence of a universal method, analyzing all risks and alternatives provides a perspective allowing us to apply methods as the situation dictates. This analysis and choice of methods, as well as the explanation of the choice, can be a visible, traceable process open for all to follow, concurrently providing credibility and flexibility. A summary of the effectiveness of the methods evaluated appears in Chapter 37.
- Full scale analysis may not be necessary in every situation. Enough clues should exist from the preliminary analysis to provide a first cut evaluation and an identification of methods which will not and which may work. Further in-depth analysis would follow for the method or methods selected. The analysis is made to fit the situation, not the reverse. Processes can be developed to enable one to carry out these approaches with minimal arbitrariness.
- Conservatism and margins of safety should not be "built in" to the analysis, but left open until cost is determined. Then, one may decide upon implementation in a visible and defensible manner.
- Methods well-grounded in theory often do not work well in practice. (We are sure this is no revelation for most of you). Uncertainties in risk-cost-benefit balancing, for example, were so great as to make this approach worthless for the cases investigated here.
- In establishing risk levels, it is possible to use *de minimus* levels below which one need not be concerned. There are many approaches to deriving these, but the one

selected should be based upon the expected impact. This means that it should be based upon risk, not upon strategies adopted to mask the risk. Implementing *de minimus* levels can conserve administrative costs for both government and industry.

- Uncertainty in health impact dominates all approaches. However, certain methods, such as the marginal cost of risk reduction and arbitrary risk levels, provide excellent guidance as to whether to adopt control over a wide range of health impact uncertainty. Again, success is situational nnot methodological.

- Analytical approaches cannot resolve value and ethical issues under any conditions. Although one can frame value judgments and problems of inequity by a particular analytical approach, the difficult judgments must still be made on less than objective grounds.

- Research could result in a better understanding of which methods work in what situations and why, as well as in a generic approach to arriving at standards in a flexible and credible manner, using minimal resources. The objective must be to resolve issues, not to continually revisit them.

III. SEQUENCE OF STUDY TASKS

As stated, the various tasks in conducting an assessment are presented, as much as possible, in the order in which they are used in the analysis. These tasks are

Task 1 — Background and Model Development — Prepare background information on industries likely to cause pollution and select or narrow choice of models to be used in the assessment.

Industries (phenol and cadmium emitting)
- Primary manufacturer or consumer
- Emitter of pollutants as by-products

Model
- River Diffusion Model (for water discharges. Corresponding models are required for other environmental media)
- Exposure Model (relating concentration in the environment to people exposed)
- Overlap Model (accounting for affects of more than one source at a particular location)

Task 2 — Health and Environmental Assessment — Determine the toxicity of phenol and cadmium from literature surveys and determine their impact on man, biota, and the environment.

Determine toxicity
- Direct toxicity
- Carcinogenic potential
- Ranges of uncertainty

Analyze environmental fate and pathways
- Environmental fate of water discharges
- Pathway analysis of water discharges
- Competing pathways and risks

Analyze errors
- Random errors
- Systematic errors
- Bias errors

Analyze policy alternatives for interpretation of data
Select a baseline set of toxicity levels and effects for subsequent analysis
Select a range of alternative levels when uncertainty is large

Task 3 — Risk Estimates — Determine releases of cadmium and phenol to water by industry to municipal discharge and directly to water. Determine ambient amounts in surface, ground, and drinking water. Use fate models and ingestion models to determine risks of each type. Do not aggregate.

Make baseline estimates based upon present exposure estimates and the toxicity baselines from Task 1. Estimate the impact of alternative assumptions to the baseline.

Task 4 — Establish Baselines for Remaining Parameters — Determine benefits of product use.

- Value of the beneficial uses of phenol and cadmium, both in dollars and other criteria by product
- Availability and costs of substitutes

Determine benefits of risk reduction

- Mortality avoided
- Morbidity avoided
- Fish and wildlife protected
- Economic loss (direct) of pollution

Determine control costs

- Costs of effluent control for given decontamination factors for each source
- Costs of cleaning drinking water both for single contaminants and for other substances
- Cost-effectiveness of risk reduced (using data from Task 2)

Task 5 — Analyze Effectiveness of Alternative Methods — Develop baseline benefits and risks. Set a "strawman" standard for each method of risk evaluation. Estimate the changes in costs, risks and benefits to society of adoption (for each item, not aggregate). Evaluate the impact of each method.

IV. GUIDELINES OF USE OF THIS STUDY

It should be re-emphasized that our purpose is to evaluate the above methods, not to definitively assess cadmium and phenol. In the absence of data, we have made assumptions (and stated them explicitly) about risks, benefits, and costs. When a specific method was sensitive to an assumption, we conducted sensitivity analyses. Secondary sources, including summaries and assessments, are sometimes cited rather than the original author since we may be more interested in the interpretation of the data in light of other factors than in the actual numbers obtained.

Prior to the actual analysis, we introduce an overview of the risk assessment and standard setting processes describing and detailing the limitation of various approaches. We trust this material will broaden your perspective on these processes. Those with an adequate background in these techniques may wish to skip to the specific analysis of interest to them.

For the broadest and most practical usage of the book, chapters are short, each containing a relatively discrete informational unit. Of course, no element stands alone, so refereences to material in other chapters is liberally provided. Chapters are organized into parts addressing broader areas. Parts I and II will assist those unfamiliar with risk analysis to grasp some of the principles used in the analysis. The parts III through X, comprising the analysis and conclusions, follow the progressive investigation of standard setting methods as applied to our examples, cadmium and phenol. Some Parts have appendices which expand upon the information presented in the chapters. We felt it necessary to provide technical background on the substance of the analysis, but did not want to encumber the reader with this information unless he was interested.

Specifically, in Part I we lay out the theoretical basis for risk analysis and discuss some of the difficulties encountered. Part II outlines the different approaches to setting acceptable risk levels on a theoretical basis. In Part III, we begin out analysis of risks posed by cadmium and phenol in water, addressing the nature, sources, and exposure routes of these pollutants and the use of models. In Part IV we discuss potency, health impact, and dose-response relationships and in Part V our baselines, including risk estimates, control costs, and estimates of the value of the technologies from which discharges arise. We also discuss suitability of alternatives to continued use of the industrial processes resulting in emissions. Part VI addresses the different types of standards available. In Part VII, we select and apply reasonable approaches to establishing acceptable levels of non-carcinogenic chronic risk from cadmium to each standard. In Parts VII and XI, respectively, we similarly examine and evaluate cancer risks from cadmium and from phenol. At no point is an actual acceptable level of risk established, since this would require value judgments. We have only evaluated the sensitivity and contribution of each approach toward aiding in arriving at such judgments. This overall evaluation and our conclusions appar in Part X.

We believe that these conclusions are adequately supported and that knowledge of this approach will make possible practical and sound standard setting.

THE PRINCIPAL AUTHOR

William D. Rowe, Ph. D. is director of the American University's Institute for Risk Analysis and professor of operations research and risk analysis in the university's Center for Technology and Administration of the College of Public and International Affairs in Washington, D. C. From 1972 to 1978, Dr. Rowe was the deputy assistant administrator for radiation programs in the Environmental Protection Agency.

Dr. Rowe is a registered professional engineer, holds 9 patents, and has published more than 250 papers and books on risk analysis and related fields.

THE EDITOR

Frederick John Hageman, M.S., has held positions in Federal regulatory agencies which have demanded an in-depth knowledge of Federal and state laws and regulations, as well as the ability to analyze complex environmental, human health, and economic problems.

His education includes a Bachelor of Science degree in entomology and plant pathology from the University of Delaware and a Master of Science degree in Science and Technology, with a specialty in environmental systems management, awarded with distinction by the American University, Washington, D.C.

He began his career as a biological aide with the Honeybee Disease Investigation Laboratory of what was then the Department of Agriculture's (USDA) Agricultural Research Service. He taught elementary and junior and senior high school biology and later served as a USDA Plant Quarantine Inspector identifying agriculturally destructive foreign insect, nematode, and plant pathenogenic pests and enforcing laws to prevent their entry into the United States.

Positions held with the Office of Pesticide Program (OPP) of the Environmental Protection Agency (EPA) have included those of entomologist, project manager, and administrative hearings coordinator.

As an entomologist, he provided scientific input to case preparation and presentation by EPA's Office of General Council during hearings to determine the appropriate regulatory actions with regard to certain pesticides. As the first OPP project manager of the Rebuttable Presumption Against Registration (RPAR) process, he coordinated, directed, and managed multi-disciplinary teams preparing scientific, economic, and legal assessments to enable regulatory decisions on products containing the pesticides chlordecone (Kepone), lindane, toxaphene, and monuron. In 1978, EPA awarded Mr. Hageman a bronze metal for commendable service for this work. He then undertook the task of coordinating and managing the OPP hearings process.

During a two-year leave of absence from EPA to work with the American University Department of Biology and the Institute for Risk Analysis (AURA), he taught undergraduate biology courses, was instrumental in initiating two off-campus programs leading to Master of Science degrees in Toxicology at industry sites and worked on AURA projects on the methodology of calculating and weighing risks and benefits. This book is an outgrowth of one of those projects.

Mr. Hageman has provided scientific input to several EPA pesticide publications, has prepared a number of position documents and Federal Register Notices and has written scripts and appeared in several of that agency's short films explaining the rationale for certain regulatory action.

CONTRIBUTING AUTHORS

Ruth Hamilton Allen, Ph.D., born in Trenton, New Jersey, obtained the degree of Doctor of Philosophy from Yale University in 1977. This degree followed three Master's degrees, all obtained at Yale, and a Bachelor's degree earned at Douglass College.

Her most recent employment, since 1979, is at The American University as Assistant Professor of Ecology and Environmental Systems Management. In this capacity, Dr. Allen teaches numerous courses at both graduate and undergraduate levels. In addition, she is responsible for planning and coordination of the Environmental Systems Management Program. She also serves as academic advisor to 35 masters degree candidates. She has been awarded course release time for program planning, course development, student recruitment, curriculum revisions, and quality control. She has served on numerous and varied university committees.

Prior to coming to American University, Dr. Allen was the principal advisor to the Chief of Environmental Analysis on Environmental matters. She was also Assistant Professor of Environmental Studies at Hood College. In addition, Dr. Allen, in the capacity of Senior Policy Advisor and Vice President of the Center for Environmental Strategy, Inc., McLean, Va., analyzed the effectiveness of national environmental policies. She also established a regional grassroots environmental organization through the Center. This followed employment as Senior Environmental Planner for the Metropolitan Washington Council of Governments where Dr. Allen planned environmental impact assessment for the Metropolitan Washington water quality management plan. In 1975, Dr. Allen was the Assistant Director of the Environmental Impact Assessment Project carried out by the Institute of Ecology, Washington, D.C. She developed interdisciplinary team analysis of environmental impact assessments. Dr. Allen also spent 3 years working at Yale University as Project Associate for a U.S. Forest Service Research Grant. She also worked as a biologist following the attainment of her Bachelors degree and while studying at Yale. Dr. Allen has received many honors and awards, notable among them an American University research award to study Delayed Childbearing, Multiple Births and Human Health, election as Secretary/Treasurer to The American University Sigma Xi Club, an AAAS/NSF Fellowship to study computerized "Information Analysis for Resource and Environmental Management," two Yale University fellowships, and an AAAS/NSF fellowship to study interdisciplinary "Biosociology".

Dr. Allen is a member of Sigma Xi, AAAS, American Sociological Association, Ecological Society of America, Water Pollution Control Federation, and other groups.

Dr. Allen has more than ten publications on such diverse subjects as risk assessment, thermal pollution, urban ecosystems, network analysis, children and nature, and NEPA. She has done more than 20 papers, presentations, lectures and performances on as wide an area of topics. Dr. Allen is also deeply involved in community and civic activities. She has been married for nearly 17 years and has three children including a set of twins.

Annette T. MacIntyre, M.B.A., is a risk analyst currently employed by Rockwell Hanford Operations, Richland, Washington. Working primarily in the Nuclear Waste Management and Disposal field, she is involved in Environmental Evaluations and Regulations. Prior to joining Rockwell in 1981, Ms. MacIntyre worked as a research assistant at the American University Institute for Risk Analysis.

Ms. MacIntyre received both her M.B.A. in Operations Analysis (1980) and her B.A. in German and Western European Affairs (1977) from the American University, Washington, D.C.

Kirk Overbey received his M.S. in Mechanical Engineering at the University of Santa Clara in 1979. While there, he did research under a multi-million dollar DOE contract investigating the feasibility of using heat alcohols as a fuel alternative. He received his M.B.A. in Finance from the American University in 1981, and worked for Dr. Rowe at the Institute for Risk Analysis. Kirk is presently working for Texas Eastern Corporation in Houston as a planning analyst. He is a member of Tau Beta Pi, Sigma Xi and the Financial Management Association honor societies.

TABLE OF CONTENTS

Part I

RISK OVERVIEW

Have I not walked without an upward
look
Of caution under stars that very well
Might not have missed me when they
shot and fell?
It was a risk I had to take — and took

Robert Frost
Bravado, 1962

In the last few years, considerable effort has been expended on developing the concept of risk assessment as well as processes and methods to effectively use the information generated. This has provided a better understanding of risks, gambles, and the means to balance and control involuntary risks. The major question outstanding is whether these risk assessment processes provide the insight and understanding which will improve control of hazardous and toxic chemicals.

The first step in preparing an answer to this question is to acquire a basic understanding of risk and of risk assessment. This step is undertaken in Part I.

Chapter 1

THE RISK ASSESSMENT PROCESS

Take calculated risks. That is quite
different from being rash.

George Smith Patton
Letter to Cadet George S. Patton IV (June 6, 1944)

Indeed, an appropriate calculation of risk is the subject of this book. Mankind has, in fact, always taken "calculated risks." It is only relatively recently that he has had the analytical techniques to weigh all relevant factors in advance for some of these risks.

I. BACKGROUND

Man has used and been subject to toxic chemicals and drugs since the dawn of history. Plant alkaloids such as belladonna, hemlock, mushrooms with both toxic and hallucinogenic properties, and grain fungi such as ergot have been part of man's lore for centuries. Developed societies such as that of Rome were subjected to lead poisoning from water pipes and food utensils and had wide knowledge of the use of many chemicals and drugs. The chronic toxicity of lead was probably unknown to the Romans, but the use of potentially hazardous chemicals to obtain benefits was widely practiced.

Since World War I, the use of industrial chemicals, pesticides, drugs, food additives for man and animal, and chemicals in other consumer products has increased spectacularly. Not only has volume of usage increased, the variety and number of different types of substances have increased as well. Many of these new substances are unknown in nature. Against them nature is virtually defenseless since their effectiveness is often due to the capability of withstanding natural deterioration and disassociation. Nevertheless, use of chemicals has led to major innovations beneficial to society.

The pursuit of these or any other benefits always involves a gamble, with risk as the downside and potential benefits to individuals and society as the payoff. This gamble is only taken if benefits are perceived greater than risks. Should the balance be favorable, one may attempt to reduce risk further while still obtaining benefits. The degree of effort and expenditure of resources for this purpose then becomes an important consideration.

Gambles of this type are often made implicitly on a voluntary basis. When those facing risks also receive benefits, and they can knowledgeably (either implicitly or explicitly) evaluate the balance, the risks are considered *voluntary*. Problems arise when the benefit and risk recipients are different. These inequitable distributions are exacerbated when the risks are imposed with neither the control nor participation of the risk-taker in the decision. Such risks are *involuntary*. Of major concern in addressing risks of toxic and hazardous chemicals are the involuntary risks imposed upon the public and the semi-voluntary risks of workers exposed to such substances.

Regulatory structures developed to assess and control risks are varied and diverse, but in the past have generally been concerned only with the risk aspect of exposure to hazardous substances. Regulators have studiously avoided considering the total societal gamble, focusing on only the limited aspect of risk assessment.

Basically, the approach has been to determine if chemical exposure will cause acute or chronic toxicity in man or animals, and if so, to restrict human exposure to "safe" levels. As long as the toxic effects were of an acute nature resulting from high exposure levels and there were known to be levels of exposure below which no effect could be measured, an approach of this type was effective. Scientists could agree on levels of toxicity, and could provide reasonable safety standards with *ample margins of safety*. These were based upon recognition of a *threshold level* below which effects could not be detected, and a margin (usually a factor of 10 to 100 lower) to account for uncertainties in both estimating risk and measuring pollutant concentration in the environment, humans, or animals. However, with increased concern for exposure to substances which may cause chronic effects, this process has broken down. It has been impossible to prove or disprove the existence of threshold levels of safety in certain toxic substances, and public health regulatory policy is presently based upon the concept of ''no-threshold'' (assuming no safe level) unless a threshold can be proven.

This approach is used primarily for substances causing the class of diseases called *cancer*. The implication is that a dose-effect response relationship exists, with doses from high levels to zero, although not necessarily in a linear manner. On this basis, data on exposure-effect relationships at high levels are extrapolated down to very low levels, using a variety of models. To date there has been no experimental validation of any such extrapolation. This includes ionizing radiation.[1] This does not mean that such extrapolations are invalid, only that validity can be neither proved nor disproved within present limitations of scientific understanding and resources.

As a result, regulatory policy for carcinogens has been a two-step process.

1.   Determining if a substance is carcinogenic in laboratory animals, and/or man
2.   If found to be carcinogenic at any level, to maximally restrict its use (within limits imposed by public acceptability determined on a political level)

For example, Section 409c(3)A of the 1958 amendments to the U.S. Food, Drug and Cosmetic Act, the Delaney Clause, states:

> That no (food) additive shall be deemed safe if it is found to induce cancer when ingested by man or animal, or if it is found, after tests which are appropriate for the evaluation of the subtlety of food additives, to induce cancer in man or animal....[2]

Thus, if a food additive is shown to cause cancer in animals, regardless of its potency, it is considered unsafe and removed from use. This applies only to those additives which have come on the market after the enactment of the 1958 amendments and not to those food ingredients ''generally recognized as safe'' (GRAS). In case one considers the Delaney Clause an anachronism, a once proposed but not promulgated rule by the U.S. Environmental Protection Agency (EPA) for hazardous chemicals, under Section 112 of the U.S. Clean Air Act as amended would have considered only the degree of evidence of carcinogenicity in any species and the potential for exposure at virtually any level in determining if a substance should be listed as hazardous. If listed, *at least* the *best available control technology* (BAT) would have been required.[3] This has been termed the ''low hurdle approach''. A different tack may well be taken in the future. Any substance found to be toxic under the U.S. Federal Water Pollution Control Act (FWPCA) amendments of 1972 (86 Stat. 816), later amended further by the Clean Water Act, required *best practical technology* (BPT) for control by 1977 and BAT 5 years later.[4] Rules once proposed by the U.S. Occupational Safety and Health Administration (OSHA) for hazardous chemicals in the workplace would have operated on the same principle. These approaches do not use the full range of available scientific and socio-political information and, as a result, may inadequately address present-day risks from toxic and hazardous materials.

To a great extent, some of these technical and information problems have been exacerbated politically, at least in the U.S. In many cases, the methods and criteria for setting standards were prescribed by a legislative body (Congress). They require regulators in the executive

---

REGULATORY LEGISLATION FOR WATER

Amendments to the Federal Water Pollution Control Act (Public Law 92-500) in 1972 set as a goal the elimination of pollution discharges, or "zero discharge," into navigable waters by 1985. An interim goal was the preservation of water quality adequate for swimming, boating, and fishing and protection of aquatic and terrestrial wildlife by 1983.[4]

All point sources, including industrial, municipal and commercial facilities were required under the National Pollution Discharge Elimination System (NPDES) to obtain a discharge permit. These permits include enforceable cut-off dates for clean-up.[4]

After 3 years of litigation and 4 lawsuits charging EPA with failure to carry out the provisions of the FWPCA, a Consent Decree was signed by the National Resources Defense Council (NRDC) and EPA, stipulating that EPA would promulgate regulations for 65 classes of toxic pollutants, including cadmium and phenol, and 34 industrial categories.[5]

In 1977, the Clean Water Act (PL 95-217) further amended the FWPCA to incorporate portions of this settlement and also set July 1, 1984 as a deadline for meeting standards.[6] As of May 1982, standards for only 2 of the industrial categories (the original 34 categories have been redivided into 24) have been issued by EPA. One of the issued standards is for the iron and steel industry.[7,8]

The administration, in May of 1982, proposed to Congress amendments to the Clean Water Act which, among other things, would extend the compliance date to 1988, not require national standards for industrial effluent released to sewers, reduce enforcement of discharge permits, exempt military bases and other government facilities from pollution requirements, double the length of a permit to 10 years, and permit EPA to assess civil penalties on violators of up to $10,000 a day and criminal fines of up to $50,000 a day *and* imprisonment up to 2 years.[9]

Those proposed amendments also requested relief from a recent court order requiring EPA to issue all effluent standards by May of 1983, asking instead for a date of July 1984.[10]

In addition to these laws requiring industrial effluent standards, the Safe Drinking Water Act of 1974 seeks to protect health to the extent feasible by using technology, treatment techniques and other means determined dy the EPA administrator to be "generally available (taking costs into consideration)." Under this Act, each state must adopt drinking water standards at least as strict as the national standards and be able to adequately monitor and enforce them.[11]

---

branch to confine activities and consideration to very narrow areas, e.g., Clean Air Act, Clean Water Act, Toxic Substance Control Act, etc. This, to some extent, has restricted the regulator to consider only those narrow issues as defined by the acts. On the other hand, the regulators themselves often take an unbalanced approach, considering only the risk element and using an extra degree of protectionism in favor of public health.

II. DEFINITIONS

In describing this process, we must define our terminology.

Risk is the chance of harm or, as previously defined, the downside of a gamble. A more formal definition is: the potential realization of unwanted consequences of an event. Both a probability of occurrence and the magnitude of the consequence are involved. Mankind has always been and will continue to be subject to risk. The concern today is primarily the involuntary imposition of risk upon members of society by technology. These risk takers, however, do not necessarily share in the benefits of the technology.

The term *hazard* implies the existence of some threat, whereas risk implies both the existence of a threat and its potential for occurrence. Thus, a threat (hazard) may exist without risk. Risk occurs only if potential pathway for exposure exists, i.e., there must be some exposure pathway to man, biota, or the environment for threat to be meaningful. In this sense, the level of exposure can be related to the likelihood (probability) of occurrence and the scope of consequence of an event.

A *toxic substance* is one for which exposure to man or animals results in deleterious effects. A *hazardous substance* is one whose effect could be of large magnitude and for which an exposure pathway may or may not exist. That is, a hazardous substance does not itself present a risk unless it has an exposure potential.

A discussion of risks and regulations may be divided into two parts: determining the risks and evaluating them. The first depends heavily on scientific input, the second more on other societal factors as well as the level of risk. The term *risk assessment* is used here to describe the total process of risk analysis, embracing the determination of both actual levels of risk and social evaluation of risks. *Risk determination*, in turn, consists of identifying risks and estimating the likelihood and magnitude of their occurrence. *Risk evaluation* measures *risk acceptance,* the acceptable levels of societal risk, and *risk aversion*, the methods of avoiding risk, as alternatives to involuntarily imposed risks. The relationship among these various aspects of risk assessment is illustrated in Figure 1. You may wish to refer to that chart while reading the definition of terms.

Risk identification and *risk estimation* are important parts of risk assessment, both involving scientific and technical determinations which usually entail considerable uncertainty. To interpret just what impact these uncertainties have, we must rely less on scientific consideration than on *judgments* by scientists and society (social value judgments).

Risk identification is of particular interest for hazardous chemicals since screening large numbers of chemicals for toxicity is costly and time consuming; especially so, given the limited availability of testing resources. Risk estimation is particularly difficult for substances which may potentially cause chronic effects, especially cancer. Major difficulties arise in relating cause and effect, and in extrapolating from high to low dose levels. Nevertheless, risk identification and estimation depend upon good scientific procedures for determining levels of risks. Conversely, evaluation of a risk depends upon certain subjective social and political decisions. This subjectivity is affected by many parameters, some of which are listed in Table 1.

A major factor in our valuation of a particular risk is whether we assume it voluntarily. Most will accept a higher degree of voluntary than of involuntary risk. Some aspects of these types of risk, i.e., degrees of the equity for the risk taken in receiving benefits of the gamble, will be discussed later.

Another factor is that we seldom discount delayed consequences of an event, such as cigarette smoking, and when we do, it is at a very low discount level. *Discounting* here refers to apportioning the risk over the years from the beginning of exposure to the manifestation of risk. We can do so for voluntary risks by translating the effect into quantitative terms where there are alternative choices as to use of money or life. However, this is not true for involuntary risks.

The *spatial distribution* of the risk in society is also important. This involves the spreading of risk from individuals to others, either in society as a whole or to designated groups. If risk recipients can be identified, our concern is greater than were we only to know that a statistical population is at risk. Then, too, one has less anxiety over the outcome of a risk if exposure to that risk is perceived as controllable.

As shown in Table 1, there are also factors affecting our subjectivity regarding risk which involve the magnitude of probability of an adverse consequence. For example, a certain level or threshold of risk is required before low probability negative consequence events are viewed with anxiety. Some risks are so low as to be considered negligible. *High probability* risks, on the other hand, are viewed quite differently, producing much higher levels of concern. The spatial distribution across society, exactly what groups are exposed, can often affect our perception of these risks. Overlying each person's view of a particular risk is an

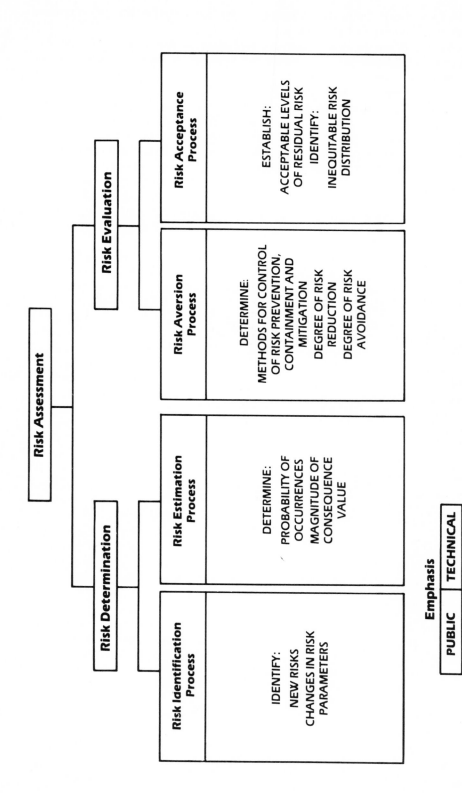

FIGURE 1. The components of risk assessment.

Table 1
FACTORS IN RISK VALUATION

Factors Involving Types of Consequences
    Voluntary and involuntary risks
    Discounting of latent risks
    Spatial distribution of risks
    Controllability of risk
Factors Involving the Magnitude of Probability that a Consequence Will Occur
    Low probability levels and thresholds
    Spatial distribution of high probability risks
    Individual risk acceptance and propensity for risk-taking
Factors Involving the Nature of Consequences
    Hierarchy of need fulfillment
    Common vs. catastrophic risks
    Special situations (i.e., military risk and life-saving functions)

individual acceptance of some risks and avoidance of others which varies from person to person.

Now, we come to our last category of factors; those which involve the nature of the consequences. One of these is the ranking of needs in a hierarchy, ranging from premature death (our greatest need in this hierarchy is the maintenance of life) to self-actualization. Obviously, consequences affecting the former need are generally perceived to be the worst. We must also consider whether a risk is "common" or catastrophic since society gives disproportionate attention to the latter which has large consequences but occurs infrequently. For example, 50,000 fatalities from auto crashes is considered a common risk since it involves many small recurring events, yet a plane crash costing 100 lives is termed a catastrophe because it is infrequent and has a high negative consequence for that single event.

There are also special situations where risks are willingly undertaken to attain a goal outside one's own personal needs. These include fighting in a war to preserve a principle or for the goals of a nation and exposing oneself to risk to save the life of another.

In addition to the factors listed in Table 1, the balancing of risks must take into account cost, benefits, and the distribution of risks in society. These decisions are as much political as social and economic. An example is the problem of regulating risk to smokers and to nonsmokers in light of the public's separation of voluntary and involuntary exposure.

Risk acceptance and *risk aversion* are interrelated and often difficult to separate. The former requires setting acceptable levels of risk. Both acceptable and unacceptable levels must be defined in the process since they are not necessarily complementary.

Risk aversion activities depend on risk acceptance and vice versa, They involve the reduction and avoidance of risk, including methods of reliability and quality control, safety research and its application, and the whole range of regulatory activities (standards, regulations, and other legal control mechanisms).

III. DIFFICULTIES IN RISK DETERMINATION

There are major difficulties in estimating risks for hazardous materials, especially in the extreme cases of accidents resulting in massive exposure and low-level exposure leading to chronic diseases. In the first instance, determining the probability of rare events is limited by lack of adequate objective historical information. In the second, it is virtually impossible to directly measure cause and effect at low exposure levels. Often, the level of health effects involved is masked by a much higher background level of ongoing effects. For example, a change in age-adjusted death rates for malignant neoplasms of 1 in 100,000 persons can be lost in the existing rate of 132 in 100,000 persons occurring from all causes. Difficulties arise because of both inadequate data bases and inherent shortcomings in the available scientific methods.

IV. SCIENCE AND JUDGMENT

Risk determination relies totally on scientific data acquisition and inference. However, science has limitations in theory, prediction, measurement, and cause and effect links. To overcome these limitations scientists must make judgments, moving the process into the area of *trans-science*. Often scientists have also been asked to assume the role of risk assessor despite the fact that science has little role in this area.

Empirical science is basically limited to designing and conducting actual experiments, speculating about the resultant effect, and developing hypotheses for explaining that effect. In the absence of supportive evidence, divergent hypotheses may/be formulated. It may be possible in most instances to determine which of these hypotheses is correct if better data is acquired. However, the cost in both time and resources often precludes generating this data, so that value judgments by scientists must be substituted.

Selection and defense of an hypothesis with residual uncertainty can be based upon nonscientific motivations and encompasses trans-science (politics, management, and business of science). The judgments involved are about science, but are not science in themselves. They are, rather, social and political determinations or, at least, show concern for social and polical realities.

Socio-political value judgments include the distribution of cost, risks, and benefits and their societal balance. A trans-scientific decision as to whether a given risk will be acceptable should nonetheless make wise use of science. This type of determination can be affected or modified by requirements to effectively administer regulatory programs. These modifications represent managerial value judgments, and are made in a different manner than socio-political assessment.

V. EVALUATION OF RISKS

Once scientific risk estimates are available, a decision must be made whether, and to what degree, to control risk. This is a social value judgment, but often may be tempered by management considerations. In general, there are three different approaches to such evaluations, with combinations thereof possible. These use as a criterion either:

1. Risk itself
2. Cost-effectiveness of risk reduction (a beneficial reduction), or
3. Balancing of costs, risks, and societal benefits

These approaches may be used to establish acceptable levels of risk for individuals or populations. Normally, various social judgments as to the degree of conservatism in protecting public health must be made. These judgments nearly always provide high margins of safety, sometimes to obtain not only a margin of protection, but also an aura of credibility in the eyes of the public and the constituency the regulator serves. An example is the language used in the Clean Water Act.[6]

Acceptable levels of risk and *unreasonable risk*, as specified in the Toxic Substance Control Act (TSCA) administered by EPA,[12] have no specific or explicit definition. However, such levels can be defined either generically, for all substances of a class, or individually for a single substance. The latter case is basically a social value judgment; the former involves more difficult judgments involving factors other than risk, such as the ease of regulatory implementation. One such case is that of *zero risk* as described in the Delaney

Clause of the U.S. Federal Food and Drug Act, administered by another Federal body, the Food and Drug Administration (FDA).[2] Absolute zero risk cannot be determined with certainty; and, in terms of risk evaluation, it means that *any* risk must be avoided or eliminated. This concept may have limited value in a specific area such as food additives, but it cannot be universally applied. Very few things are risk free; eliminating one risk may directly or indirectly raise others. You can see that overall perspective is necessary for generic risk estimates.

The consequences of harmful materials may include morbidity, premature death, and property damage. However, the cause of death, the manner of dying and injury, and the type of risk involved all influence the subjective evaluation of these consequences. The causes and manner of dying fall into four general classes:

1. *Immediate acute effects* — Immediate death or injury from a specific event involving explosion, fire, suffocation, corrosives, poisons, etc.
2. *Delayed acute effects* — Delayed death or injury resulting from massive toxic exposure as a result of a special event or set of events
3. *Observable chronic effects* — Premature death or increased morbidity resulting from exposure to identified substances over a long period. The effects are often cumulative as in metal and pesticide poisoning or latent as in carcinogenesis
4. *Unobservable chronic effects* — A contribution to premature death or increased morbidity by the hypothesized synergistic or contributing action of a particular substance. Some substances indicate toxicity at high levels of exposure, but their action at low levels cannot be established. Others operate synergistically, as in the exposure of smokers to asbestos

Society reacts to these consequences in different ways since the value an individual places on each is primarily subjective.

Two questions must be addressed by the risk assessor in the valuation of the consequences of disease and accident. First, is death or illness from disease in general different from death and injury from accidents and, if so, how? Second, are death and illness from different diseases valued differently, and if so, how? Intertwined with these questions is the extent of the ability of society to control the progress of the disease through diagnosis and therapy.

To answer the first question, disease vs. accidental consequences, we must take into account a number of factors such as the degree and duration of pain and suffering, lingering vs. sudden death, known disease progression, decreased quality of life as a result of illness or injury, and so forth. With the majority of accidents, there is an initial shock and trauma; then, assuming the victim is still alive, a period of recuperation that often ends in complete recovery. This process can occur with heart attacks. However, with most disease processes, there is a time lag between onset, diagnosis, and therapy which is not usually associated with traumatic injuries. Many chronic diseases with limited cures have a reverse process involving the initiation and recognition of the disease when the patient is in good health, followed by a progressive decrease in health until death, arrest of the disease, or a full or partial cure. Such degradation, and the fear of it, can have enormous psychological and physiological effects. On this basis, cancer is one of the most dreaded diseases. Of course, accidents can result in maiming and permanent disability, but do not elicit the same degree of dread.

Although many diseases are not fatal, they may be debilitating, involve pain and discomfort, and affect the quality of life and the pursuit of one's livelihood. The perceived need, the legal requirement to grant sick pay, and the loss of the productive capacity of workers make such illness an industrial problem.

A heart attack may be a sudden and unexpected event that is shocking to the victim's family and friends, but is absent the same kind of deterioration that accompanies cancer.[13] Diseases, such as cancer, which bear a long-term negative prognosis with little hope of cure or remission, are perceived as being of great negative impact. Conversely, some of the most dreaded diseases in history, such as leprosy, can now be cured or arrested and are therefore no longer feared as much as they were in the past. As a result, answers to our second question depend on both the ability to cure a disease and the interval between onset and death.

Dividing the risk assessing process into risk determination and risk evaluation, discussed previously, is to some extent artificial. The degree of uncertainty in risk determination always affects how we evaluate risks. The degree of evidence of carcinogenicity, for example, affects how a substance is evaluated and controlled. In general, however, it is the large uncertainty in risk estimation itself that leads to major problems in the evaluation and regulation of toxic substances.

Chapter 2

CRITICAL REGULATORY DECISIONS

Still one thing more, fellow citizens — a wise and
frugal government, which shall restrain men from
injuring one another, which shall leave them
otherwise free to regulate their own
pursuits of industry and
improvement, and shall not take
from the mouth of labor the bread it has
earned. This is the sum of good
government, and this is necessary
to close the circle of our felicities.

Thomas Jefferson
First Inaugural Address
(March 4, 1801)

Therein lies the crux of the problem. How do we restrain men from injuring others via toxic pollutants while still leaving them otherwise free to regulate themselves and not impairing the fruits of their endeavor? Our founding fathers, no doubt, could not imagine the complexities faced today in applying their axioms to the problems of a technologically advanced society.

Regulatory agencies are faced with a series of critical decisions with respect to both new and existing chemical compounds. With limited resources, they decide not only whether a substance in commerce is hazardous, but also how to address untested and new substances. These decisions must be based both on the available scientific information and regulatory objectives. Balancing gains and losses, selecting methods to do so, and establishing a method for enforcing all involve social and managerial decisions.

Although specific to benzene use in the U.S., the following excerpt from a *Washington Post* editorial[14] provides a broad overview of the problems facing regulators of hazardous chemicals:

> If you are responsible for ensuring health and safety in the work place, what do you do about a substance that is known to cause cancer, but for which the quantitative relationships — how much is dangerous and over what period of time — are unknown and will likely remain unknown for many years? Do you set the lowest exposure standard that is technically and economically feasible — even if meeting that standard is quite costly? Or, in the face of so much uncertainty, do you leave in place a higher — and riskier — but generally accepted standard while the search for better evidence continues? If you choose to impose the more stringent standard, must you be able to demonstrate a numerical relationship between the cost of meeting the standard and the number of lives that are likely to be saved? Or is it enough to know that lowering exposure will save lives even if you don't know how many? In short, how do you balance cost, lives and uncertainty?
>
> These are the questions with which the Supreme Court must grapple in its consideration of a case that could have profound implications for the ways in which this country adapts itself to living with toxic chemicals. It is a case that rests heavily on the testimony and knowledge of scientific experts and crystallizes the dilemma facing scientists who are thrust into policy debates. For a scientist faced with so much uncertainty, the course of action is clear. He does not publish. His experiment is not over until further research reduces the uncertainty to an insignificant level. He goes back to the lab and collects the answers he needs. But the policy-maker, and the scientists he depends on for advice, have no such luxury. They have responsibilities, mandated by law, which force them to act.

I. IS A SUBSTANCE TOXIC OR HAZARDOUS TO HUMANS?

Substances are hazardous as a result of inherent flammability, corrosibility, explosiveness, toxicity, or combinations of these attributes. Deciding how to deal with these chemicals often depends upon the choice of control to obtain a particular level of risk, where it is otherwise inefficient to use resources for either regulation or accident prevention for lower levels of risk. For accidents, the direct measures of resulting events provide a feedback mechanism to continuously monitor the need to change regulatory policy.

Evidence of direct cause-effect relationships of toxic substances and human health effects are tenuous or nonexistent. The degree of validity of this evidence from tests on laboratory animals depends upon the type of information acquired and the specifics of the experiment. Tests of animals similar to man in certain aspects of biological make-up are probably more significant than are tests of laboratory rodents, but may still be difficult to relate to man. Even epidemiological evidence, probably the most valid evidence for human toxicity, depends upon the strength of the study design and the ability to statistically relate cause and effect.

Mutagenicity tests at the bacterial or cellular level, such as the Ames test,* provide evidence or suspicion of carcinogenicity. This may also be suggested by chemical and pharmacological data on known carcinogenic compounds with structures similar to substances under consideration.

While uncertainty exists for all evidence, the judgment of scientists is used to bridge this information gap. This often requires extrapolation from animal effects and from high laboratory dose levels to human response at low dose levels. Two different approaches are

1. Estimating the most likely extrapolation model
2. Selecting a model with a built-in safety margin

The first relies upon scientific judgment and available evidence; the second, a social value judgment as to the safety margin desired. The latter is only valid after the former has been established, i.e., the safety margin should not be set by scientists, but left as a social regulatory judgment. This seems appropriate since one must consider trade-offs among the cost of safety, the degree of protection afforded, and the value to the regulatory process of increased credibility.

II. HOW SHOULD WE TREAT NEW OR UNTESTED SUBSTANCES?

Strategies to deal with new or untested substances must factor in availability of suitable validated testing resources, who will bear the cost of testing, and the disposition of the substance during the test period. Strategies to prioritize tests are usually based upon potential threat to the public or to workers. This threat involves the potential toxicity and exposure levels by use, as well as the ability to control the use. One strategy is the use of a *screening mechanism*. Categorization of screened substances would be based upon factors such as: how the substance concentrates in tissues or organs; the intake level of the population at risk (including high intake groups); toxic and carcinogenic potential of metabolites; results of both long- and short-term tests, chemical structure, and prospective use conditions.

Who should conduct the tests? Should it be industry, the regulatory agency, or designated laboratories? How can such tests be validated and replicated? How can trade secrets be protected? Who bears the cost of testing? These are some of the more obvious questions that a regulatory strategy must address. More subtle questions include whether use should

---

* The Ames test is a standardized study using a mutated strain of salmonella bacteria to determine whether introduction of a given substance causes further mutations in the bacteria. Because mutagenesis is likely to be a prelude to carcinogenicity, the test is used as a low-cost indication of the latter. There has, in fact, been a very high correlation of positive Ames response and carcinogenic response in laboratory animal tests for most tested substances.[15]

be restricted during the test period, what is the impact of premature release of information to the news media, and what are the barriers to innovation established by the high cost and business risk of entry of new products.

A restriction to existing substances, as opposed to those in test phases, implies that a risk-balance is made in each case, because removing a product from commerce might increase risks in other parts of society, e.g., restriction of pesticides used to combat fruit flies during a regional outbreak, or saccharin for control of weight and diabetic balance. Loss of this balance is in addition to other benefits foregone by removing the product from the market. On this basis, independently examining each case on its own merits might be better than an across-the-board policy. Guidelines for determining the conditions for imposing restrictions would enable establishment of procedures balancing regulatory flexibility and the action prescribed.

There are still other sources of error. For example, the premature release of test information allows misinterpretation. One reason for this is that tests designed to examine dose-response under a specified condition may not be valid for other conditions. For example, in a study of rats exposed to high levels of formaldehyde through inhalation pathways (to simulate extreme human occupational conditions), several of those sacrificed were found to have developed nasal tumors. This cannot be related to the general human population since the primary exposure pathway here is ingestion at very low levels, but it certainly does not imply that formaldehyde might not be toxic through low level ingestion. (Use of formaldehyde-based insulation can result in nonoccupational inhalation). It only means that the published data has no direct bearing on toxicity for this pathway.

Once people become apprehensive, significant effort is necessary to reverse the process. Public anxiety is initiated when new risks are identified, and seldom quantified. This lack of quantification precludes a risk assessment prepared in a manner in which a *risk agent** can compare the new risk with other risks. Anxiety cannot be alleviated until the scope, magnitude, and ability to take risk-aversive action are known. Thus, a new risk may be magnified by lack of knowledge. If quantified, a risk might have a lower value than perceived. Furthermore, subsequent quantification does not always change the public's preconditioning caused by a qualitative risk estimate. The initial impression, erroneous or otherwise, is often difficult to reverse. This may in turn lead to the dilemma of choosing between ascertaining the propriety of early reporting of an identified risk and waiting for adequate quantification.

Early reporting may be the only viable option since it may be the only way to focus attention and resources in a manner that allows later quantification. Failure to report new risks prior to quantification may lead to charges of "cover-up" and loss of credibility.[13] Therefore, data should be made available to authorities promptly and completely despite the possibility of misuse which could erect barriers to the flow of information.

The high cost of toxicological testing, estimated to be in the millions of dollars for some cases, and the possibility of resulting restrictions raise the cost of entry into the market place. This not only reduces competition to large organizations with adequate resources, but may stifle innovation. When individuals and small organizations cannot afford to experiment, new substances whose potential benefits are yet unknown may never be pursued because of suspected toxicity. However, even if a substance is found to be highly toxic, its end use may result in very limited exposure and, hence, low risk.

The high cost of bioassays to determine carcinogenic potential requires that most testing be conducted at very high doses, well above ordinary exposure levels, to determine effects from a relatively small population of laboratory animals. Extrapolations are then made to lower doses, even when evidence is sketchy that these lower doses produce cancer. A common criticism of this process is that high doses overwhelm the organism's ordinary detoxification and cellular regeneration mechanisms.

---

* Individual at risk.

III. WHICH SUBSTANCES SHOULD WE REGULATE AND CONTROL?

This decision depends upon the level of risk expected from use. This means that pathways to humans must be known and that the substance or its metabolites must be observable in organs and tissues. Detectable levels must exist in both the environment and in man (in vivo) with adequate resolution for both the level of risk and type of information needed to establish that risk.

Establishing an *acceptable risk level** requires a social value judgment. However, this judgment is not the only consideration, since the ability to regulate is also important. One must ask: is regulatory control to take place at the ambient level, the emission or the effluent source, or all of these? Can all sources be controlled? Can regulations be enforced?

IV. ENFORCEMENT

Because a chain of evidence is needed for legal purposes, enforcement ability depends upon whether compliance can be monitored. This requires measurement capability at a given location(s), either permanently or temporarily, with technical staff adequate to obtain and assure validity of the information. The performance, maintenance, and calibration of measurement equipment and the performance of personnel must be carefully controlled, documented, and witnessed.

Since the resources necessary to exhaustively accomplish this task are very large, trade-offs are required between the degree of voluntary and of involuntary compliance. This raises the following questions: how much enforcement is actually needed to encourage voluntary compliance? To what extent can regulatory authorities depend upon data obtained and submitted by those regulated? What form of regulations are best enforced? Is enforcement equitable? None of these questions is easily answered, but they must be since enforceability is a major factor in every decision to regulate.

V. INTERRELATIONSHIPS AMONG CRITICAL DECISIONS

All of the critical risk decisions are interrelated to some extent. Two specific cases are worth emphasizing here. The first is made in light of the fact that a particular regulatory framework can focus scientific research in narrow, sometimes unfruitful, areas. We must be aware that selection of a particular test method requires concentrating limited resources on improving that method at the expense of new approaches which in the long run may be more sensitive and less expensive. The second is the recognition that regulation sometimes is necessary in the face of both scientific and socio-political sources of uncertainty. The immediacy of need for control and protection of the public vs. the costs of wrong or poor decisions is a constantly faced dilemma. The trade-offs between prescribed procedures and flexibility to handle special cases are dynamic. While science can help reduce the uncertainty in risk estimation, it cannot be expected to do away with the need for social trade-offs. Science, even good science, has its limitations.

VI. NATIONAL IMPLICATIONS

As indicated in the *Washington Post* editorial cited earlier, the major question is how to regulate in the face of scientific uncertainty. There are two separate, but related approaches:

1. Developing strategies for better determination of risks given limited resources

---

* Any term such as unreasonable risk, unacceptable risk, etc., may be used in place of acceptable risk here. The problem is semantic, not methodological.

2.   Developing approaches for resolving the problem of balancing risk, cost, and benefits

Judgments made in each area reflect on the other.

Each country has its own regulatory structure and profusion of regulatory laws. In some, including the U.S., responsibilities are fragmented, diverse, and in some cases overlapping; in others, responsibility and laws are more focused. The scope and effectiveness of such laws and regulatory policies is not within the purview of this book. Except for illustration, we will not explore specific regulations, regulatory structure, or cases. We will, however, discuss the underlying problems and approaches to these problems.

VII. INTERNATIONAL IMPLICATIONS

While one must assume that the appropriate authorities in each nation establish their own regulatory structure and regulations based upon balances suitable for their society, some problems transcend national concern. Two of these are the international trade of hazardous materials and the movement of pollution across national borders.

Because of differing regulations, international trade is often hampered by import-export restrictions on regulated and controlled chemicals, drugs, and toxic substances. Should requirements be different for control of substances for use within a country as opposed to those for export? Some examples may illustrate this problem. The U.S. forbids the use of DDT within its borders (except for specific cases involving public health), but many un-developed countries find benefits of use outweigh the risks. Should the U.S. allow its manufacture and export?

There are other situations where products or chemicals are based in one country and permitted in another. Certain drugs used in Europe are restricted in the U.S. Some European countries have put restrictions on the use of "Glow Lites" for night fishing because of the large amounts of tritium involved. Should we? Trade in building materials using asbestos poses the same kind of problem.

One must also consider the global impact of the use of toxic chemicals within a nation which may cause pollution in other countries. Acid rain in Scandinavia, chemicals in the Danube or the Great Lakes, and contamination from runaway oil wells, such as in the Gulf of Mexico, are examples.

Obviously, efforts to achieve consistency in the control of toxic substances is important, especially so in the areas of laboratory practice, chemical nomenclature, and key terms relating to control and data requirements for new toxic substances in international commerce. One solution is the pooling of data to provide broader bases. This includes information on good laboratory practice and standardization of experimental design, protocols, and reporting techniques, as well as on methods to validate data and protect trade secrets.

Chapter 3

DETERMINING RISK

And everyone that heareth these sayings of mine, and
doeth them not, shall be likened unto a foolish man,
which built his house upon the sand:
And the rain descended, and the
floods came, and the winds blew, and
beat upon that house, and it fell:
and great was the fall of it.

Jesus
Matthew 7:24-27 (KJV)

Any determination of risk, to be credible and accurate, must be based on a strong foundation. This requires an orderly approach and a proper consideration of factors which, if unheeded, could greatly affect the final result.

I. STEPS IN CONDUCTING THE STUDY

There are three separate steps in conducting a study such as undertaken here. The first is establishing a *baseline* against which we can evaluate different methods of setting risk levels in standards. The second is setting those levels by a number of different methods using available data, including data from the analysis used to establish the baseline. The last step is evaluating each method by comparing the results of each with the baseline. This provides an estimate of the relative impact of each type of method. A number of critical parameters are useful to make these comparisons.

We considered the following parameters:

1. *Health effects estimates* — estimates of the exposure and toxicity of populations at risk via the environmental medium (air, water, etc.) in question. Where different estimates of risk or exposure were critical, multiple estimates were made
2. *Environmental impact* — estimates of impact to the biotic and abiotic components of the environment at the present level of exposure
3. *State of technological utilization* — estimates of value of business and other benefits as a result of use of the substance in question
4. *Costs and performance of control* — estimates of the present costs of controls and their performance capability. Direct costs are the only ones considered here

In all cases, the measures used should be the most appropriate for the category or subcategory considered. We have not aggregated these into a single measure of utility or value. In the last step, each method of establishing risk levels resulted in changes to each of the four measured parameters in relation to the baseline. The relative importance and value of each parameter is left for the reader to determine, i.e., we made no value judgments of this type.

II. TREATMENT OF ERROR AND UNCERTAINTY

At the outset, one must consider basic sources of uncertainty and errors and how they propagate. We have addressed these problems by using relative risk approaches with *ranges* of alternatives.

A. Study Error Propagation

We use relative (or comparative risk) approaches avoiding absolute risk estimates to the extent possible.

Environmental effects, use, commerce, health, and increased costs of control are the measurement parameters. The change in these parameters against the baseline provides a common set of relative scales to use for evaluation. Errors in numerical estimates due to inaccurately estimating measurement parameters are methodologically the same for all methods so all scale up or down together. Determining the cost-effectiveness of the different approaches and balancing risks, costs, and benefits are useful as surrogate measures, but are subject to absolute errors.

B. Uncertainty

Treatment of uncertainty can lead to other potential sources of error in the analysis. Where there are wide ranges of uncertainty, one can make multiple estimates to test the sensitivity of the evaluation methods. This is particularly important in estimating health impact, less so in economic areas. In the latter case, errors in our ability to measure predominate, while in the former, judgment is required as to the correct way to estimate risk in the absence of information. These process errors are more significant than the measurement error.

III. TYPES OF RISKS

A single substance may produce a number of different consequences. For example, smoking not only causes lung cancer, but the available evidence indicates an involvement in heart disease, emphysema, and chronic bronchitis. This necessitates consideration of the total impact of risk.

A. Acute Effects

Some hazardous chemicals cause a variety of effects for which cause-effect relationships can be established. Although the effects may be latent, that is, there is a delay between the time of exposure and the onset of effects, the causes are usually distinct events either causing exposure or involving continual exposure at significant levels. Although there is no exact definition, these threats are usually termed acute. Some examples of acute threats are events such as accidents in the transport of toxic materials, misapplication of pesticides and herbicides, industrial accidents, potential sabotage or terrorist acts (such as the release of toxins to water supply or food) and continuous exposure to high levels of toxic chemicals (such as that of fishermen in Minamata, Japan to mercury and of workers producing chlordecone (Kepone®) in the U.S.).

For most substances, there exist *threshold* levels above which acute effects are manifested. Such thresholds may be caused by a variety of conditions. For example, the body may have the capacity to eliminate or metabolize certain levels of intake but, if these intake levels are increased above the elimination capability, bioaccumulation may occur until a threshold is reached. Alternatively, there may be little or no metabolic accommodation. Sometimes, the absence of trace nutrient metals such as copper may add to the deleterious effect.

Acute effects are not usually detectable below a threshold, i.e., a *No Observable Effect Level* (NOEL). Establishing a NOEL is a standard toxicologic procedure used both here and abroad when safety factors are employed. Generally, a standard is set to prevent build-up to a threshold, either by keeping exposure limits well below elimination and metabolic capacity or lifetime accumulation limits within *ample margins of safety*.[16] The Environmental Protection Agency (EPA) states, ''the NOEL is defined to be the level (quantity) of a substance administered to a group of experimental animals at which those effects observed or measured at high levels are absent and at which no significant differences between the group of animals exposed to the quantity and an unexposed group of control animals maintained under identical conditions is produced.''[17] Safety factors are added to keep actual levels well below the threshold. For example, for food additives the *acceptable daily intake* (ADI), or tolerance level, is obtained by dividing the NOEL by 100; this is based on the rule of thumb that man may be more sensitive by tenfold than the experimental animal used and that there may, in addition, be a tenfold variation in individual sensitivity. This ADI applies for all regulated food additives except those which are potentially carcinogenic.[16]

Note that setting a NOEL involves risk estimation and that setting margins of safety (to establish ADIs) requires a consideration of uncertainties and evaluation of risk in terms of the *degree of safety*. One should not confuse these processes.

One aspect of the uncertainties in establishing levels of risk in man involves varying sensitivity in human populations for dose-effect relationships. These are difficult to establish since epidemiological studies are restricted to *targets of opportunity** (for ethical reasons) and extrapolation of animal data provides little information about man. Problems in measuring and evaluating test results are additional aspects that further contribute to the difficulty in estimating risks and establishing threshold levels where they exist.

B. Chronic Effects

As discussed under ''Acute Effects'', a few chronic effects such as heavy metal poisoning and high level exposure to pesticides (e.g., dioxin) and other industrial chemicals can often be directly related to the cause since measurable quantities of toxin can be detected either in the environment or in tissue. However, sometimes there are latency periods between exposure and effect, multiple exposures from varied sources, or synergistic responses. Carcinogenesis, mutagenesis, and teratogenesis are principally of this type. Cause and effect are difficult to relate for these diseases, especially at low exposure levels.

Although problems with chemicals and compounds for which directly observable cause-effect relationships are important, the most difficult area of control and management is for those substances which cause *stochastic* effects, i.e., effects which are probabilistic as a result of exposure. Carcinogenesis is of primary concern in this area.

Toxic materials often affect specific organs. The effects of mercury and lead on the nervous system, cadmium on the kidney and lungs, and certain mushroom toxins on the liver are all well documented. Others are systemic, causing or enhancing the incidence of cancer and tumor growth in many organs or causing more than one disease.

The difficulties in estimating risk of chronic disease from low-level exposure to toxins arise from both the inability to obtain an adequate data base and inherent shortcomings in the available scientific methods of relating cause and effect. The latter is especially true in environments with competing and masking effects. These are discussed in detail in Appendix I-A.

---

* A person or persons, who, under normal living or occupational conditions, is (are) subjected to a particular harmful agent.

Chapter 4

LIMITS FOR RISK ESTIMATION

We may, of course, strike a balance between what a living
organism takes in as nourishment and what it gives out in
excretions; but the results would be mere statistics
incapable of throwing light on the inmost
phenomena of nutrition in living beings.
According to a Dutch chemist's phrase, this would
be like trying to tell what happens inside a
house by watching what goes in by the door and
what comes out by the chimney.

Claude Bernard
Introduction a l'Etude de la Medecine Experimentale (1865)
(translated from the French)

Similar to nutrition in M. Bernard's time, human response to toxins can be difficult to determine even today. For example, precisely what happens? What organs are affected? How severe is the impairment? Our inability to answer all such questions with total confidence imposes limits for risk estimation.

The risk evaluation process is sensitive to variation in the degree of confidence one has in risk estimates. On this basis, we have investigated a number of ways of obtaining dose-effect and exposure measures. The resulting range of possible values was kept and used in evaluating standard setting methods. Once this was done, we assessed the sensitivity of each method to both the variation and the degree of evidence.

I. ACUTE EFFECTS AND NOEL

A high intake of a toxic chemical can affect many organs resulting in a variety of diseases. There may be a multiplicity of no observable effect levels, NOELs, for each disease and organ afflicted.

Unless one has data on acute toxicity in humans for a particular chemical such as those existing for cadmium and mercury, toxicity determination must be extrapolated from animal tests. When doing so, the choice of model is critical. For example, for ingestion the effects in animals may be related to humans by either equivalent dietary exposure continued for a lifetime or by daily intake as a fraction of body weight averaged over a lifetime. The latter is usually expressed as milligram intake per kilogram body weight per day. Since rats and mice daily consume a larger fraction of their body weight than do humans, this is less conservative than the first approach by a factor of between 2 1/4 to 4 for rat data (depending upon the strain) and a factor of 6 for mouse data.[18] For inhalation models, the different air intake to lung mass ratio, among other things, must be considered.

If one established a lifetime NOEL, daily intakes could be derived via appropriate routes based upon actual exposure patterns of recipients. Intake and metabolic processes for converting or eliminating toxic materials on a short-term basis must be balanced to enable one to project whether the body burden will surpass the NOEL.

When several data sources present different results or bases for comparison, we must evaluate the experiments to determine their scientific validity. There are many criteria which can be used, but the difficulty of assigning quantitative factors to the quality of information precludes a sophisticated analytical treatment. Reasonable judgments, however, can be used to rank studies in terms of quality or, if not, at least to determine which experiments are valid.

The estimate uncertainty due to measurement uncertainty, model uncertainty in extrapolation, and variation of species sensitivity are all confounding factors, not easily quantified. Therefore, we use ranges of estimates spanning these uncertainties.

II. CHRONIC EXPOSURE

Chronic exposure estimates have all of the problems mentioned in the preceding section, as well as some of their own which exacerbate the risk estimation process. First, the toxin concentration is below that at which acute effects occur. At these levels, one often encounters the need for long exposure periods and the occurrence of latency periods between initiation of stress and effects. Secondly, the absence of NOELs or the inability to establish them means that we must expect effects in diminished form at very low doses, perhaps even to the zero exposure level. Since testing cannot provide significant answers at these levels, we must extrapolate from higher exposure levels. The choice of models to accomplish this is wide, e.g., one-hit to multi-hit models; our findings are exceedingly dependent upon our selection of models.

There are three types of estimates which may be pertinent to chronic exposure models:

1. Degree of evidence of effects
2. Estimate of relative potency
3. Extrapolation to low doses

A. Degree of Evidence

First one must review experiments to decide whether they provide adequate evidence of chronic toxic effects in test species and, perhaps, humans. All effects must be considered. For example, if benign tumors result from exposure to a substance, this occurs irrespective of whether there are other tumors which may be malignant. In carcinogenesis, benign tumors may or may not be precursors to malignancy — that judgment must be addressed on merit or through other experiments for each situation.*

The results of analysis, often based on selecting the most "reliable" test results, are dichotomous — producing either no-effect or some level of effect. Thus, the analysis should focus on the degree of evidence that an effect occurs.

B. Relative Potency

As we have said, one method of classifying the potency of a given substance is to relate it to that of other chemicals subjected to the same type tests. This provides a means to associate classes of chemicals by their general potency level, obviating the need for accurate measures of absolute values, which are difficult or, in many cases, impossible to derive.

Potency is usually expressed in units which are the inverse of units of administered dose or of environmental exposure. Thus, if the substance is a food additive, concentration by weight is used to express the dose rate. Table 2 shows the different units used for various pathways of administration of toxins. The determination of potency, especially in humans, is difficult and, as previously discussed, involves large ranges of uncertainty.

In order to develop a model based upon *relative potency*, we will need a baseline for comparison. One possible reference level, arbitrarily established, is equivalent to the parts per billion of exposure via the relevant pathway which would, over the average life of an individual member of a species, increase mortality by one in a million.

An alternate model uses the risk of loss of life expectancy as opposed to lifetime risk. Some substances, such as 2-Acetylaminofluorine (2-AAF) depend upon time and rate of

---

* For this reason EPA refers to tumor-producing substances as *oncogens* (tumor producers) rather than carcinogens.

Table 2
POTENCY UNIT FOR SOME DIFFERENT PATHWAYS AND DOSE

Pathway	Daily dose rate	Method of calculation	Concentration	Units-potency
Food (ingestion)	$mgkg^{-1}day^{-1}$	Weight	1 ppm	$mg^{-1}kg$ day
Water (ingestion)	$mgb^{-1}day^{-1}$	Weight	1 ppm	$mg^{-1}day$
Air (inhalation)	$ppm\ day^{-1}$ $mgm^{-3}day$	Volume Weight/volume	1 ppm .81 ppb[a] .001 ppb[b]	$ppm^{-1}\ day$ $mg^{-1}m^3day$

[a] By weight.
[b] By volume.

Table 3
POSSIBLE RELATIVE POTENCY REFERENCE LEVELS

	Dose concentration	Risk increase
Human-animal lifetime risk model	1 ppb/day	1 in 10^6/lifetime
Human-animal annual risk model	1 ppb/day	1 in 1.5×10^8/ year
Human-animal life expectancy model	1 ppb/day	143 sec (2 1/2 min) of average life expectancy

exposure so that short term impact may mask longer term effects at lower dose levels.[19] Therefore, change in the average life expectancy by administered dose, as shown in Table 3, may be a better measure than, for instance, increased numbers of tumors. This is especially true when the time span of dose is taken into account. *Cohort* analyses for estimating changes in life expectancy are now available,[20] but even this measure is incomplete in that morbidity and quality of life are not taken into account.

Table 4 is an example of a ranking for some substances where dose is administered through ingestion.[22,23] The estimated central value in parts per billion necessary to achieve a one in a million lifetime risk is listed under the second column. The relative potency shown in column (a) is the reciprocal of the values on the left, rounded to one significant figure. These figures are further rounded in column (b) to the nearest order of magnitude, consistent with the general degree of uncertainty about the potency. Column (c) is an integer index derived from the exponent of column (b).

On an arbitrary basis, chemicals can be classified by the relative potency scale (column c) into any number of groups. We propose three as an illustration:

1. High potency ≥ 0
2. Moderate potency <0, but ≥ -3
3. Low potency < -3

Ethylene thiourea or DDT provides a baseline to separate high and moderate potency,

ORGAN ACTIVITY

When considering potency of ingested substances, one must keep in mind that the liver is the most important site of metabolization or alteration of the toxin. Occasionally, the metabolite is more toxic than the original substance. The liver puts the altered substances into the blood or bile; constantly receiving recirculated blood and thus has many opportunities to alter substances. Of course, metals such as cadmium are elements and are not reducible.

Inhaled pollutants may be absorbed into the bloodstream at the sites of oxygen exchange in the lungs, the alveoli. They may then be transported to organs throughout the body, often initially bypassing the liver.

Although the liver is the major site of biotransformation, the MES (microsomal enzyme system) of cells, where the reactions occur, are also active in the skin (where dermal absorbtion can occur), the gastrointestinal tract, and the respiratory tract.[21]

CONCENTRATIONS

It is interesting to view the chemical concentrations we will be discussing in terms of everyday concepts and items. This has been done by Dr. Warren B. Crumett of the Dow Chemical Company in the following table:

Trace Concentration Units

Unit	1 Part per million	1 Part per billion	1 Part per trillion
Length	1 in./16 mi	1 in./16,000 mi	1 in./16,000,000 mi (a 6-in. leap on a journey to the sun)
Time	1 min/2 years	1 sec/32 years	1 sec/320 centuries or 0.06 sec since the birth of Jesus Christ
Money	1¢/$10,000	1¢/$10,000,000	1¢/$10,000,000,000
Weight	1 oz/31 tons	1 pinch salt/10 tons potato chips	1 pinch salt/10,000 tons potato chips
Volume	1 drop vermouth/80 "fifths" gin	1 drop vermouth/500 barrels gin	1 drop of vermouth in a pool of gin covering the area of a football field — 43 ft deep
Area	1 ft²/23 acres	1 in.²/160-acre farm	1 ft² in the state of Indiana or 1 large grain of sand on the surface of Daytona Beach
Action	1 bogey/120 golf tournaments	1 bogey/120,000 golf tournaments	1 bogey/120,000,000 golf tournaments
	1 lob/1,200 tennis matches	1 lob/1,200,000 tennis matches	1 lob/1,200,000,000 tennis matches
Quality	1 bad apple/2,000 barrels	1 bad apple/2,000,000 barrels	1 bad apple/2,000,000,000 barrels
Rate	1 dented fender/10 car lifetimes	1 dented fender/10,000 car lifetimes	1 dented fender/10,000,000 car lifetimes

and benzo-α-pyrene a possible baseline for separation of moderate or low potency. This case is presented only to illustrate a relatively imprecise but useful classification scheme.

C. Extrapolating to Low Doses

When designing a study, it must be of adequate duration and have sufficient animals to produce usable results. Because cancer is so devastating, a low incidence in the human population is important. Therefore, animals must be abundant enough and dose high enough

Table 4
RELATIVE RISK MODEL BASED UPON POTENCY

Substance (ingested)	ppb[a]	Relative risk level[b]		
		(a)	(b)	(c)
Dimethylnitrosomine (FDA)	0.05	20	10^1	1
DDT (FDA)	0.4	3	10^0	0
Ethylene thiourea (FDA)	2.0	0.5	10^{-0}	−0
NTA (FDA)	260	0.004	10^{-3}	−3
Vinyl chloride (FDA)	7	0.1	10^{-1}	−1
Aflatoxin (Wilson)	.0002	5,000	10^3	3
Benzo-α-pyrene (Wilson)	700	.001	10^{-3}	−3
Saccharin (Wilson)	6×10^5	.000002	10^{-6}	−6

[a]   Parts per billion for a lifetime risk of approximately one in a million.
[b]   Column (a) is to one significant figure; column (b) is to the nearest order of magnitude; column (c) illustrates classification by the exponents of column (b).

so that a low incidence can be seen. Strains with a high natural incidence of cancer are also frequently selected. It is in the extrapolation of this data to man that arguments arise.

There are a number of different models for extrapolating from high to low doses for continuous dose-response functions. The most cited is the linear model, although concave upward models are used to account for changes in response due to dose rate, and convex upwards models are sometimes used to account for highly sensitive members of a population. The overall model may be a logistic curve (S-curve) which is linear at low dose, of higher degree at higher doses, and finally saturates at very high doses where the effect in question, such as mortality, always occurs.

Frequently, in practical problems many models fit the data equally well, although they yield substantially different extrapolations at the usually lower doses of interest (Appendix I-B describes these models in some detail). To find the effects of different models on extrapolations made to low doses, or to estimate doses needed to produce low response levels, Schniederman has taken four different mathematical models (Figure 2) and fitted these to animal data (from Maltoni) on the effects of vinyl chloride.[24] Aiming for a response of 10^{-8}, given the dose level which produced a 1 in 100 response rate, the fractions of that dose required by each of the models were four:

1.   One-hit model — 10^{-6}
2.   Logit (slope: -3.454)10^{-4}
3.   Probit (Mantel-Bryan slope: 1)10^{-3} to 10^{-4}
4.   Two-hit model — 10^{-3}

The point is that, no matter how much testing occurs, reduction of the three orders of magnitude of uncertainty at the extrapolated risk level is unlikely. The basic uncertainty may be irreducible on an absolute risk basis. Thus, one may have to use relative risk.

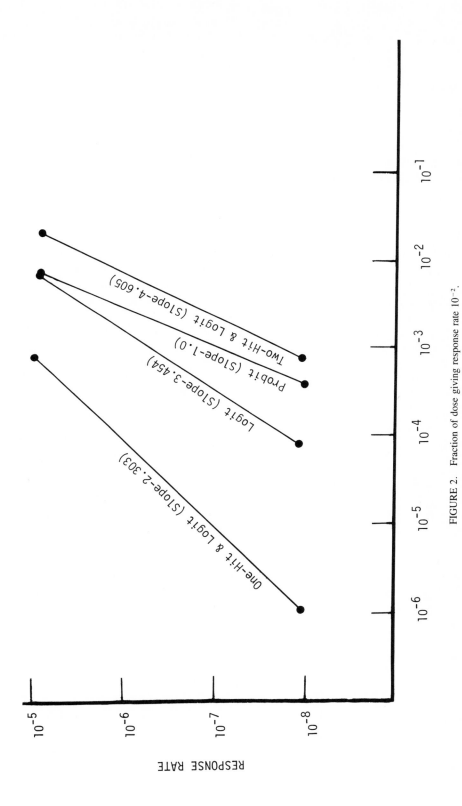

FIGURE 2. Fraction of dose giving response rate 10^{-2}.

Appendix I–A

MAJOR PROBLEMS IN DETERMINING CHRONIC RISKS

Major problems determining chronic risks may be divided into four sometimes overlapping areas:

1.   Data bases
2.   Limits of scientific approaches
3.   Data acquisition
4.   Experimental design and interpretation

I. DATA BASES

Developing data basis of low level epidemiological and animal studies to determine cause–effect relationships is fraught with difficulty. Data for epidemiological studies are derived primarily from health records and vital statistics, since subjecting humans to potential threats for experimental purposes, either on a voluntary or involuntary basis, is viewed as unethical in most societies. Targets of opportunity provide the basis for most data. These studies are most often retrospective, i.e., an after–the–fact assessment. In these situations, relatively few parameters are either known with any degree of precision or under control of the experimenters. For example, data on individual exposure, as opposed to average group estimates, are usually nonexistent. In addition, mortality and morbidity records may not accurately record the cause of death or even provide the proper diagnosis of illness. There are economic, political, and social reasons for this. Some contributing factors are the cost in time and money of health professionals in keeping accurate records or obfuscating the cause of death for protection of a family, insurance purposes, or protection against third party liability. The lack of accurate recording is abetted by constitutional guarantees against invasion of privacy in some countries, including the U.S., which make it impossible to record even the basic data required. Uncontrollable parameters, such as mobility in the population; other stresses such as smoking, which mask effects in question, and cultural and social practices, such as high fat diets, all add to uncertainty. In general, because only gross statistics are obtainable, only relatively high–level cause–effect relationships can be found.

Animal studies present similar problems. There are, for instance, variations in animals within species from different or even from the same environments, inappropriate lifetimes for long latency periods, difficulty in cross–species correlations, and differences in experimental design. Trade secrecy and differing regulatory data requirements are examples of institutional problems. As a result, there is, at present, no means of normalizing data obtained in one set of experiments for comparison with that from another set by different experimenters under different circumstances.

II. LIMITS OF SCIENTIFIC APPROACHES

Even if data are adequate, one may be limited by resources, inherent uncertainties, and the inability to control all important experimental parameters. More importantly, extrapolation, whether across species (animal data to man), within species (sensitivity to stresses), or from high to low dose-response, is inherent in nearly all interpretation of experimental results and poses additional limitations.

To determine dose/effect, either epidemiological or animal laboratory studies may be used. With regard to the former, there are several types.[21] These are

- *Retrospective* — locate a group with a disease suspected of being caused by risk factor. Use questionnaires, interviews or records to reconstruct exposure situation. Compare with a control group
- *Prospective* — select a healthy group and gather information on them over a period of years. Compare those exposed to those not (control group)
- *Retrospective/geographical* — assess the given siuation by geographical region

There are a number of problems involved in using any of these approaches. They include the fact that the *effects* of the disease may be falsely associated with a toxin, the subjects could forget effects or exposure, or, by having a disease, could become more aware of risk factors and exaggerate their importance. The prospective studies, because of the latency factor, become very hard to follow and subjects or controls for one reason or another can drop out and their records become lost. Often, investigators can better track the living than obtain information on the dead, which can skew results.

If a free choice is offered to subjects or controls or if one or the other group is different in significant variables, another major error is introduced. For example, one cannot compare exposed workers to the general population where some are incapable of working. All factors considered, epidemiological studies cannot provide the final word but can indicate the possibility of responsibility by a toxin.*

In epidemiology studies, the basic design may often be questioned. Biases may exist in the population samples as well as with the experimenters. When questionnaires or interviews are used to determine population history, potential exposure, or habits, the manner in which the questions are posed, or even the questions themselves, may affect the results. Adequate design requires factorial experiments to minimize variation due to the questions. *Blind* and *double blind designs*** can be used to minimize the bias of experimenters and those under study.

Basic limitations in such studies also arise from uncontrolled parameters such as changing populations, limits to longitudinal data acquisition, competing threats to the population, identification of valid control groups, and basic random variations in measurement.

Resource limitations can also adversely affect the experiment. These can restrict the size of the experiment, the scope of data obtained, the length of time the experiment is carried out, and its integrity in light of annual budget perturbations. In many cases, taking into account the sample sizes required for statistical verification and the ability to control parameters during the experiment preclude undertaking a study since at the outset only inconclusive results can be expected.[25]

With regard to animal studies, the arguments opposing reliance on them, many with a degree of validity, have been repeated many times:[21]

1. There are so many problems in rearing the animals, preventing unusual stress, maintaining accurate records, selecting appropriate numbers and dose levels, and setting up good controls (usually litter mates and same sex as the subjects) that a high degree of error is introduced
2. Man is unique and animal responses do not duplicate ours (this is best minimized by

---

* For an exceptionally clear presentation of advantages and disadvantages of animal and epidemiologic studies see Reference 21.

** A blind design is one in which the subjects and controls have no knowledge of what is being tested and why. A double blind design study extends that ''blindness'' to the data gatherer as well.

choosing an animal with a target organ system with a response as close to man as possible)

3. Too high a dose is usually given. More than one or two doses should be administered
4. Tests are usually not of adequate duration to determine chronic effects (this can be overcome with generation studies)
5. Synergistic or antagonistic effects of other chemicals cannot possibly be measured with current technology
6. Unless "double blind" studies are used, the investigator often applies his or her own biases

Animal studies, especially those involving lifetime testing in intact animals, are expensive, slow, and not particularly sensitive. When completed, all they provide are conclusions about cause-effect relations in the species studied at the levels of exposure involved. Studies at multiple levels of exposure are used to extrapolate downward to lower dose levels, but such extrapolation is a value judgment as is the expectation that man will behave in a metabolically similar fashion to animals. Even as one proceeds up the phylogenic scale from mouse to great ape, there may be greater likelihood of similarity of response, but each case and organ studied may be different. Another difficulty occurs when animal tests are carried into the period where normal old-age lesions begin to occur. In many such cases, the data are confounded, and it is impossible to construct an unbiased statistical test.[26]

Another source of evidence is the chemical structure of compounds which we can use to predict the metabolic fate and potential toxicity of substances. The following quotation from the 1975 monograph of the National Academy of Science-National Research Council (NAS-NRC) illustrates the wide acknowledgment of this fact:

> Many important decisions, at least about the sequence of testing, can be made without testing at all on the basis of analogies with other known chemicals. Structure-activity relations are reasonably well understood for some groups of chemicals and some toxic effects, less well-known for others. However, many new industrial chemicals differ only trivially from other known materials and relatively few fall into genuinely unknown groups. Those that do will require correspondingly more complex testing.[27]

There are four classes of compounds with regard to the levels and conditions of use:

Class 1. Compounds of simple organic structure that are readily handled through known metabolic pathways and without adverse biochemical, physiological, or pharmacological effect

Class 2. Compounds structurally analogous to those in class 1 and whose metabolic fate can reasonably be assumed to be unassociated with any adverse biochemical, physiological, or pharmacological effects

Class 3. Compounds with structures so different from those in classes 1 and 2 that reasonable assumptions regarding metabolic fate and freedom from the possibility of adverse effect are precluded[28]

Class 4. Compounds with structures similar to known carcinogens and carcinogenic metabolites and which would be expected to have similar effects

Such evidence is not as strong as that from human and animal studies, but it does provide significant additional information.

Another indication is the short-term test. These tests are now available at the mammalian cellular level for the identification and study of substances which present a possible cancer hazard.[29] In recent years, a number of systems testing neoplastic cell transformation by chemical and physical carcinogenic agents have been developed. Some of these are being used in several different laboratories with good reproducibility, and others are still in the development stage. The most widely studied are the golden hamster embryo cell system and the mouse embryo fibroblast cell line systems.

In these neoplastic transformation tests, cells derived from transformed colonies or foci, when innoculated into syngeneic* or immunosuppressed animals, can grow as malignant tumors. While the definitive evidence for neoplastic transformation of cells in culture remains their tumorogenicity in animals, a number of phenotypic changes of the cultured target cells are commonly used as indicators.

Other in vivo systems using specialized cell types are currently being developed. Among these are systems using epithelial cells from liver, epidermis, and other organs. Neoplastic transformation by chemicals of well-defined epithelial cells have been achieved in vitro; conditions for quantitative studies are under development. Such systems may be needed to identify particular critical target cell populations within target tissues closely correlated with carcinogenesis in vivo.[30]

For an effect to occur, most chemical carcinogens require metabolic activation by cellular enzymes to an ultimate reactive metabolite. In mammalian species, this carcinogenic activation takes place in many different organs and tissues. Cells in culture can retain enzymatic activity, but specific culture systems or preparations may lack or lose the necessary enzyme for activating a particular chemical. Therefore, we should give adequate consideration to the effectiveness of metabolic activation functions in each test system. In the absence of animal bioassay and epidemiology data, short-term tests for chemical carcinogens presently do not constitute definitive evidence of adverse human effects. They can, however, provide a clue as to whether a substance poses a carcinogenic hazard to humans.

Data on metabolic pathways, enzymatic action, and synergism are a precursor to further evaluation and testing.

III. DATA ACQUISITION

Resource and logistic problems associated with toxicological studies of laboratory animals pose perplexing problems. For instance, manageable sample populations cannot detect increases in toxic events of less than 5 to 10%. For some health effects such as mutagenesis and teratogenesis, incidences in the human population of 3 per 1,000 or 3 per 10,000 are significant. Obviously, these health effects have not been well studied in whole-body mammalian assays where only the strongest of such effects could be observed. The magnitude of this problem is illustrated by the fact that in assays for environmental toxicants using 1000 animals, each animal is a surrogate for 200,000 people in the U.S.[31]

Problems of scale occur in epidemiological studies as well. For example, to detect a doubling of a cancer risk, if 10% of the Los Angeles population were exposed, one would need to study at least 570 exposed people. For a fivefold increase in risk, a baseline of 40 would be required. To put this in perspective, only 1,000 new cases of squamous cell carcinoma of the lung occur annually among the entire white male population of Los Angeles County.[32]

In fact, surveys rarely encounter counties where as many as 10% of the residents are employed by any particular industry and thus exposed to toxic substances used in that industry. Generally, no more than 1 to 2% of the population works in any one industry. Thus, a 7% increase in lung cancer, if actually due to the occupational exposure of the small percentage of the population employed in that industry, indicates a risk to workers four to eight times that of the general population.[33] County increases of only one or two cases per 100,000 people can be quite significant given the relatively small proportion of workers exposed.[34]

There are limited testing resources for both epidemiological and animal studies throughout the world. Many facilities do not use standardized and validated test protocols, preventing

---

* Animals with identical genotypes, as maternal litter mates.

effective data sharing. There is, in addition, great variability of data base quality which often makes pooling of data difficult if not impossible. Therefore, there is a great need to standardize data acquisition methods.

IV. EXPERIMENTAL DESIGN AND INTERPRETATION

As Davis (1979)[34] has pointed out:

> . . . public health now reflects a complex interaction between environment, lifestyle and genetic factors. Toxicological studies of single pollutants necessarily exclude such confounding factors. Epidemiological studies of human populations can embrace some of these multiple factors, but it is rarely possible to devise an epidemiological study which can confirm clear causality — that exposure to substance X causes Y disease, under Z conditions.
>
> Lave and Seskin (1977) discuss what an *ideal* study of air pollution and mortality might entail. They note that genetic factors are now difficult to measure conceptually, while data on lifestyle factors such as smoking and nutrition do not generally exist or are poorly measured.

One must first estimate the level of detection desired, and then design the experiment so that significant results can be achieved through selected protocols. A checklist with at least the following considerations is mandatory:

- Identify controlled and uncontrolled parameters
- Determine length of the experiment and the ability to control changes over time
- Decide on the confidence level to be used
- Determine the type of statistical distribution expected (or use a normal distribution)
- Statistically design the test for: size of samples, number and size of control samples, criteria for evaluation of positive and negative outcomes
- Develop bias control protocols: open, blind, double blind
- Evaluate the data using proper statistical techniques
- Make comparisons and define the methods chosen where choice of statistical technique can affect the results

After the experiment is completed, we must interpret the data. Speculative inquiry and scientific judgment now become important, and the line between science and trans-science becomes very fine.

Appendix I-B

EXTRAPOLATION MODELS

There are several alternative models used in extrapolating for carcinogenic effects. These are:

I. LINEAR OR ONE-HIT MODEL

The dose response model which has probably received the most attention in analyzing dichotomous data is the *one-hit model*. The basis for this model is the concept that the response can be induced after a susceptible target has been hit once by a single biologically effective unit of dose. This implies that the probability of response p_d at dose d is given by:

$$p_d = 1 - \exp(-\gamma d) \tag{1}$$

where γ is an unknown parameter. For small values of d (i.e., low dose), it follows that (approximately):

$$p_d = \gamma d \tag{2}$$

For low dose levels, this linear model is essentially numerically identical to the one-hit model. The parameter γ represents the slope of the linear dose-response relationship.

For carcinogenesis, there is no adequate experimental data on the shape of dose-response curves at low levels (for example, that below a response of 1 in 100). It is, however, currently assumed that most dose-response curves are concave upward in the low dose regions when dose is plotted against response. Intuitively then, the linear model provides an upper bound to such dose-response curves and hence a conservative estimate of that dose which yields a probability of response below a specified level.[35]

II. PROBIT, LOGISTIC, AND EXTREME-VALUE MODELS

For any cumulative distribution function, F, a whole class of models is given by:

$$p_d = F(\alpha + \beta \log d) \tag{3}$$

where α and β are parameters whose values must be determined. If F is the cumulative normal distribution function, then the model states that the probit of the probability of response is linear in log dose. The logistic and normal distributions have a long tradition in the analysis of quantal response data. This in itself, however, does not justify their use in extrapolation. The probit model is related to the use of a lognormal distribution in time-to-occurrence models. When we use the Weibull distribution in place of the lognormal, we obtain the extreme value distribution for quantal response.[36] The three models can be written as:

$$\text{Probit: } p_d = \Phi(\alpha_1 + \beta_1 \log d) \tag{4}$$

$$\text{Logistic: } p_d = [\exp(\alpha_2 + \beta_2 \log d) + 1]^{-1} \tag{5}$$

$$\text{Extreme value: } p_d = 1 - \exp[-\exp(\alpha_3 + \beta_3 \log d)] \tag{6}$$

where Φ denotes the cumulative normal distribution function. For $\beta_3 = 1$, the extreme value model becomes the one-hit model and is therefore a one-more-parameter generalization of the one-hit model. For each of these three models, incidence data may be used to directly estimate the unknown parameters α and β. Furthermore, we can construct a dose estimate at any prescribed confidence level to be less than or equal to the unknown dose level associated with any particular desired response. This is the standard statistical approach to the problem. The difficulty is that it assumes we can realize the unknown dose-response curve within the functional form used in the estimation procedure. Although not critical for interpolation problems, this assumption is extremely critical for extrapolation some distance from the experimental range.[35]

III. MULTI-STAGE MODELS

The *multistage models* are natural extensions of the one-hit model.[37-40]

In simplistic terms, the biological rationale for the multistage model is that there are (K) biological stages which the chemical must pass through (e.g., metabolism, covalent binding, DNA repair, etc.) without being deactivated, before the expression of a tumor is possible. We assume that each stage, (i), is linear with dose; response i = i + Vi dose and

$$\text{Risk} = 1 - e \prod_{i=1}^{-k} (\alpha i + Vi) \tag{7}$$

The multistage models will still be conservative in their estimation of risk. For instance, in the specific case of two experimental doses, $K = 2$, V_1, $V_2 = 0$, the model will *always* overestimate the risk at the lowest dose level (assuming $K = 1$ has been rejected statistically), and will approach linearity at low doses.[41]

IV. GAMMA MULTI-HIT MODEL

Another model which one should consider as an alternative to the one-hit model is the *gamma multi-hit* class of models.[39] The risk at low dose ∂ is

$$= \int_0^{\lambda\partial} \frac{\mu^{k-1} e^{-\mu}}{(k-1)!} \, \partial\mu \tag{8}$$

$$\text{RISK} = \sum_{x=k}^{\infty} \frac{e^{-\lambda\partial} (\lambda\partial)^x}{x!}$$

where λ and k are estimated parameters.

For small integer values of k and low doses, this model can be shown to be similar to the multistage model defined earlier. For integer values of k, the above integral has a closed-form solution including a polynomial in dose with fixed powers of λ as coefficients. Although this model is less general than the multistage model, the conceptual and mathematical derivations can be framed in such a way as to be similar.[41]

V. OTHER MODELS

In addition to those we have discussed, there are a variety of other models, such as the Weibull distribution and those that take into account latency, competing risk, and life-shortening which we will not describe here.

Part II

ESTABLISHING LEVELS OF RISK IN STANDARDS

For us, with the rule of right and wrong given us by Christ,
there is nothing for which we have no standard. And
there is no greatness where there is not simplicity,
goodness and truth.

Leo Tolstoi
War and Peace (1865-1869)
pt. XIV, 18

A standard, according to Random House Dictionary, is anything serving as a rule for making judgments or as a basis for comparison. A second definition is anything authorized as the measure of quantity or quality.

The second definition is, of course, that of a Federal or state imposed standard discussed in these pages. The first definition, however, is relevant in that the standard selected must be sound enough to serve as a stable point from which one may accurately compare and judge other levels.

We will later evaluate a number of different kinds of standards for control of cadmium and phenol in water. In each of these a wide range of methods of setting acceptable levels of risk, including those presently in use, are examined. The four basic risk evaluation approaches to setting acceptable levels of risk are based upon:

- Risk aversion practices
- Establishment of acceptable levels of risk
- Economic or related techniques
- Particular regulatory strategies

There are, however, a number of philosophical questions underlying any approach to risk estimation and evaluation which cannot be resolved without addressing ethical concerns.

In carrying out tests, as discussed in Part I, there is always a trade-off between quality of data and risks of testing. Testing of human subjects (unethical because of high risk) would, of course, produce the best data, but is not permitted. There are also many who believe animal tests to be unethical even though without them the quality of available data would be severely limited.

Given populations have distributions of sensitivity to particular stresses which may or may not be genetically inherited. A major question we must address is the extent to which we should protect these sensitive members by broad regulations. Cost to the general population may be excessive if we set standards to protect its most sensitive members. The basic issue is, if sensitive individuals may be identified in advance, should they exercise a greater degree of protection and, if so, should this be voluntary? There are no easy answers, but the question cannot be ignored.

A final problem is whether and how to equitably distribute residual risks and general societal benefits. Those who take the risks often may not receive any benefits. To what extent should such inequities be minimized? The answer must be found on ethical and political, not scientific, bases.

This part contains a general discussion of some of the advantages and limitations of each method of setting levels of risk for standards.

Chapter 5

ESTABLISHING LEVELS OF STANDARDS BASED ON RISK PARAMETERS

Look to your health; and if you have it, praise God,
and value it next to a good conscience; for health
is the second blessing that we mortals are
capable of; a blessing that money cannot
buy.

Izaak Walton
The Compleat Angler
Part 1, Ch. 21

Obviously, good health is an important commodity. So much so that some methods of setting standards are based solely upon adverse effects to human health. In this chapter, we discuss and assess such methods.

I. RISK AVERSION METHODS

All risk aversion methods are based upon the *a priori* idea that any risks are undesirable and should be avoided. They preclude a discussion of benefits which may otherwise offset the risks.

A major value judgment underlies all approaches, namely, where to stop. There is no absolute measure, even for *zero risk*, only a reference to conditions. Nevertheless, the direction of value judgments for these approaches tends to be as conservative as possible in order to foster protection of public health.

A. Attaining "Zero Risk"

A zero risk concept, i.e., intolerance of any known level of risk, is *global* in theory, but can only be locally applied in practice, i.e., the removal of risk of one specific (local) type does not necessarily result in the lowering of societal risk (global). Further, since one can never be assured that every risk has been identified, actual zero risk is impossible.

To bring a risk to zero (or very low level) locally, i.e., for a given risk exposure pathway, one must eliminate or avoid the underlying activity, find zero risk substitutes, or apply absolute controls. In applying absolute controls, substances or practices with risk thresholds can achieve zero risk on an absolute basis without complete cessation of the practice. The controls must be set below the threshold with a margin of safety to provide assurances in the face of uncertainty and variation in measurement. However, where no threshold can be demonstrated, some risk occurs at any exposure level. Thus, absolute control is not possible in these cases unless we eliminate the source of exposure. This can, in turn, raise risks in other pathways. An example is the prevention of industrial discharge of an amount of a substance which would produce chronic effects. This could result in accumulating wastes to a degree that exposure becomes high enough to cause acute effects at the waste locations. The chronic exposure to the population before this high concentration may, on balance, have been less risky than the new threat to individuals.

Generally, the concept of zero risk applies to local processes. The overall process which provides a global approach included risk-balancing of all affected risks. The Delaney Clause of the Federal Food and Drug Act as amended[2] is an example of a "zero risk" approach for food additives. As in the case for saccharin, it has been demonstrated to be a local approach.

B. Risk-Balancing

Reduction in one risk may cause increased risk elsewhere. As a result, all risks resulting from a change of control strategy should be considered *globally* to allow the total risk to be balanced. However, there is no single scale available for making such balances. In addition, when there are different kinds of risks to different populations, one must address perceptions of severity of consequences, equity, and a host of other parameters in the process of risk balancing. This approach is useful in identifying the kinds and ranges of affected risks, but not in guaranteeing any unique balance. It does, however, provide perspective on value judgments that we might need to strike an arbitrary balance.

A specific case is the use of substances to preventively block the toxicity of other substances. For example, zinc, calcium, and absorbic acid can block the toxicity of cadmium while normal potassium iodide is used to prevent absorption of radioactive iodine (^{131}I, ^{129}I) in the thyroid.[42,43] Of course, the risk of intaking the prophylactic substance must be less than that of the substance which t protects against.

C. As Low As Can Be Measured

A variant of the zero risk approach is to reduce the source of risk to the lowest measurable level. This acknowledges the imprecision and inaccuracy of measurement, the inability to enforce control if residues cannot be measured, and the need of obtaining a legal chain of evidence. Note that measurement capabilities in the laboratory are very different from those for field measurements under actual environmental conditions. Tests in one mode cannot be easily replicated in the other.

A related problem is how to handle confounding processes such as the *background* level from either similar causes or similar and competing effects. Since background effects often have significant variation in occurrence and measurement, this background variation may become a limiting process.

D. As Low As Can Be Controlled

The *as low as can be controlled* approach refers to the inability to control contributing factors on a local basis, not to the technical control capability. For example, when considering control of cadmium in water, the existing levels retained in silt and river bottom, which may have been deposited from air or natural sources, are dominating, uncontrollable factors. The cadmium is retained in inorganic and dead organic matter,[44] is sensitive to pH level, and is often at a concentration orders of magnitude above that in solution.

When utilizing this approach, one should recognize both the existence of uncontrollable parameters and the fact that parameters should not necessarily be controlled just because they can be. Although it is essential to consider the ease and relative value of this control before imposition, this is often not done.

E. As Low As Background

The *as low as background* approach is a variant of the *as low as controllable* approach in that the uncontrollable parameters are natural background causes or effects. This can be a major problem when investigating cancer in man or animals because of the incidence of spontaneous cancer of both age-related and nonage-related types. Variation in background becomes significant for measurenent and for the meaning and value of control at very low levels.

II. ACCEPTABLE RISK APPROACHES

Acceptable risk approaches are based upon the concept that there are levels of risk one must assume involuntarily for the opportunity to live in a highly developed society.

URANIUM MINERS AND FLORIDA HOMEOWNERS

A good example is the problem of measuring elevated levels of radon gas in houses built on phosphate land in Florida.[45] In this case, background levels of radon are at 0.004 WL (working level) — the unit describing any concentration of short-lived decay products of radon-222 in one liter of air, which results in the release of 1.3×10^5 MeV of potential alpha energy. The limit of measurement is at 0.005 WL. The variance of neither measure is given. The proposed guidance is set at 0.015 WL, including background, less than twice the range of uncertainty in measurement and an example of the *as low as can be measured* approach. Although direct measurement is not possible at this level, one can extrapolate from higher doses to uranium miners.

These data for the uranium miners are among the best epidemiological data available for chronic effects of any type; nevertheless, argument exists even here as to interpretation. EPA estimates that air exposure of 0.02 WL, 75% of the time over 70 years would result in 2000 excess lung cancer deaths in a 100,000 cohort. However, uncertainty in measurement without evaluating variation in background and in health impact makes a standard set this low quite controversial. EPA, in proposing such guidance, did take into account a number of other factors, including the cost of control. This is particularly crucial since it may be the homeowner who will pay for remedial action.

Terms such as *acceptable levels of risk* and *unreasonable risk* cannot be defined explicitly, but their existence is evident from experience. Determining them is a social process involving the balancing of cost, risks, and benefits. Distribution of these elements is often inequitable. Recognizing difficulties in quantifying social variables and the impossibility of unanimous agreement on any social issue is a prerequisite for understanding the factors that make a level of risk acceptable.

Unquestionably, some risks are acceptable. Under certain conditions, acceptance is evident, such as when:

- A risk is perceived to be so small that it can be ignored — threshold condition
- A risk is uncontrollable or unavoidable without major disruption of lifestyle — status quo condition
- A credible organization with responsibility for health and safety has, through due process, established an acceptable risk level — regulatory condition
- An historic level of risk continues to be acceptable — de facto condition, and
- A risk is deemed worth the benefits by a risktaker — voluntary balance condition

There have been many definitions of terms such as acceptable, unacceptable, and unreasonable risk. Most definitions encompass the methods of establishing references so that risk may be compared. These references include different ways preferences may be expressed and evaluated.

A. Risk Comparison

To determine acceptable risk levels, we compare the investigated risks of both individuals and populations to benchmarks, criteria, or value judgments. Table 5 lists ways of establishing risk acceptance levels. Note that the maximum level of risk may be different for individuals, groups, or society as a whole. This is true both for the measurement of risk and the type of risk.

1. Measurement of Risk

We must determine whether to use objective or perceived risk in estimating actual risks.

Table 5
RISK APPROACHES USING COMPARISONS TO OTHER RISK

Maximum acceptable level of risk to individuals, groups, or society
  Measurement of risk
    Objective estimates of risk
    Perceived risk
      Revealed preferences
      Implied preferences
      Expressed preferences
  Types of risk
    Voluntary vs. involuntary risks
    Catastrophic vs. ordinary risks
    Immediate vs. latent risks
    Controllable vs. uncontrollable risks
Risk trade-offs
  Risk brokering — redistribution without a change in overall risk
  Average risk to society vs. most exposed individual
  Statistical vs. identifiable risks
  Global risk trade-offs — reduction in one risk may increase risks elsewhere

The dichotomy arises from two sources, the uncertainty in measurement and the variability in individual perception. In the first case, the estimated risk cannot be fully equated with the actual because estimates of the probability and consequence of occurrence comprising a risk estimate may be inexact. Variability in *risk perception* is affected by factors which have been explored by many investigators[13,46-48]. Considerable effort is, in fact, still underway to evaluate these factors.

One may make *risk comparisons* for perceived risks; obtaining references for acceptable risk by three different methods: revealed, implied, and expressed preferences.

1. *Revealed preferences* — This method is based on the assumption that, by trial and error, society has arrived at a nearly optimal balance between the risks and the benefits associated with any activity. One may, therefore, use statistical cost, risk, and benefit data to reveal patterns of acceptable risk-benefit trade-offs. Acceptable risk for a new technology is assumed to be the level of safety associated with ongoing activities having similar benefit to society.

2. *Expressed Preferences* — The most straightforward method of determining what people find acceptable is to ask them. The appeal of this method, the expressed preference, is obvious. It elicits current preferences, thus being responsive to changing values. It also allows for widespread citizen involvement in decision-making and, therefore should be politically acceptable. It has, however, some possible drawbacks which seem to have greatly restricted its use. For example, people may not really know what they want, their attitudes and behavior may be inconsistent, their values may change so rapidly as to make systematic planning impossible, they may not understand how their preferences will translate into policy, they may want things that are unobtainable in reality, and different ways of phrasing the same question may elicit different preferences.[49]

3. *Implied Preferences* — The implied preference method may be seen as a compromise between the revealed and expressed methods. It looks to the legal legacy of a society as a reflection of both what people want and what current economic arrangements allow them to have. Its proponents, like those of the democratic process, make no claims to perfection; rather, they see it as a best possible way of muddling through the task of bringing risk management in line with people's desires. Its problems are familiar to any participant in a democracy. Our legal legacy includes not just laws adopted by our elected representatives, but also interpretations and improvisations by

judges, juries, regulators, and others. Laws are sometimes inconsistent and poorly written, often extended to cover situations undreamed of when they were written, with precise formulation often reflecting fleeting political coalitions or public concerns.

2. Types of Risk

There are a number of factors regarding a particular risk which have a great impact on the way the risk is perceived and regulations imposed. These include whether the risk is voluntary or involuntary, of "great" or "small" magnitude, of immediate or latent impact, and controllable or uncontrollable by available measures. All of these factors will be discussed as we evaluate methods of establishing risk levels for our selected pollutants.

3. Risk Trade-Offs

The value judgments made for acceptable levels of risk alone are also affected by trade-offs among risks. Risk equalization and/or brokering seeks to spread the risk equitably and are forms of risk balancing. The objective is to minimize exposure to particular individuals or groups without necessarily reducing overall levels of risk. For example, lowering population risks from radiation at nuclear reactors may require higher exposure to occupational workers, perhaps providing a limit on risk reduction without resorting to industry shut-down as an option. One also must decide whether to attempt to reduce risk to the most highly exposed, the most sensitive, or the "average" citizen and whether that reduction is to be based upon statistically defined or readily identifiable risk. In this latter case, we are more concerned with risks to people whom we can explicitly identify than for general members of a population, especially if the population at risk is remotely located from us.

B. Arbitrary Risk Numbers

The objective of the methods described in the preceding section is to obtain reference levels to use as benchmarks for making value judgments about limiting specific risks. One researcher believes levels of acceptable maximum risk to individuals can be established universally on an annual basis, such as an involuntary risk of one part in a million for the public and ten parts in a million for a worker.[50]

A once-proposed Environmental Protection Agency (EPA) Clean Air Act standard for hazardous pollutants and the currently proposed Federal Drug Administration (FDA) standards for "Sensitivity of Method" for animal food additives consider one in a million lifetime risk as an absolute number for decision-making. In these cases, the purpose is to indicate where it is "safe" to stop acquiring information on possible risks.

The difficulty of setting acceptable levels of risk by considering risk alone is not only confounded by inequitable distribution, but also by the size, age, and general health of the exposed populations. Risk to an identified risktaker is different from that to a statistical member of the population. In addition, large risks to a few people and small risks to large numbers of people are not directly reconcilable. Concern for both aspects requires dichotomous approaches. At best, such approaches seem to be more appropriate when used as benchmarks for comparison where they provide guidance in addressing a wide variety of regulatory questions, such as effective use of limited resources, than when used to derive rigorous numerical standards. Once made binding, they lose much of their value to stimulate questions and may become prescriptive without merit.

An *arbitrary risk number* is similar, but not quite the same as a *de minimus* level. In the first case, we assume that the risk number is small enough to be below the societal benefits of all activities. The *de minimus* level, which considers only risks independent of any benefits, assumes that the risks are trivial enough to be beneath consideration of the law. Thus, *de minimus* levels are legal conditions either too small to be worthy of the attention of the courts or too low to measure or enforce.

In Chapter 6 we address economic approaches to risk control. Included among these is cost-effectiveness which requires establishing arbitrary risk values in terms of the amount spent to avoid a risk. When a specific value for reduction of a particular health effect is chosen across society, it becomes an arbitrary risk number (or set value for cost of risk reduction) to the same extent as does a risk number alone. In other words, a value is assigned below which risks are not controlled. An example is the use by the Nuclear Regulatory Commission (NRC) of $1000 per man-rem in their "Safety Goals for Nuclear Power Plants."[51] Other approaches may be used to establish such values, but on a less universal basis.

C. Technology Based Approaches

Technology based approaches are those which rely upon the practical reduction of risk by available control technology. They assume a degree of acceptable risk. The *"bubble concept"*, which will be described, is a method of allowing an overall level of acceptable risk by total effluent from a single plant.

1. BAT and BPT

The Clean Water Act distinguishes between new and existing plants in its requirements. *New Source Performance Standards* (NSPS) must use Best Available Technology (BAT) under Section 306 of the Act since industry has an opportunity to design and apply the best and most efficient technologies for new plants. New plants which indirectly discharge pollutants, i.e., discharge to publicly owned treatment works (POTWs), must also use BAT, for the same reason, for *Pretreatment Standards for New Sources* (PSNS).[6]

Pretreatment Standards for Existing Sources (PSES), likewise, are to be analogous to BAT for removal of toxic materials which pass through, interfere with, or are otherwise incompatible with POTWs. Cadmium and some other toxic metals may pass through and be deposited in sludge, limiting its agricultural use, or find its way via water to man.[6]

Federal guidelines for state effluent limitations for a number of industrial categories will include the degree of reduction achieved by BAT and Best Possible Technology (BPT).

2. The "Bubble Concept"

The *bubble concept* has primarily been used in air pollution control where overall risk levels for pollutants are established within a given area and individual sources are permitted to trade off pollution within the limit.

The bubble concept has recently been extended by EPA to water effluent guidelines and standards. Here, the concept applies to an individual plant. Dischargers with multiple outfalls may discharge greater amounts of pollutants from outfalls where there are high treatment costs in exchange for an equivalent decrease at outfalls where abatement is less expensive. This would allow discharge of no more total pounds of pollutants but should give flexibility to allocate discharge most cost-effectively.[8]

Existing National Pollutant Discharge Elimination Systems (NPDES) permits may be modified to reflect use of the bubble concept at any time during the life of that permit if the concept was not considered when issuing the permit.[8]

Chapter 6

ESTABLISHING LEVELS OF STANDARD SETTING BASED ON BOTH RISK AND BENEFITS

I took one Draught of Life–
I'll tell you what I paid-
Precisely an existence-
The market price, they said.
They weighed me, Dust by Dust-
They balanced Film with Film,
Then handed me my Being's worth-
A single Dram of Heaven.

Emily Dickinson, in the Complete Poems
of Emily Dickinson, edited by
Thomas H. Johnson (1960)
No. 1725 (n.d.)

Just what is a life worth, one might ask. Such a question must be addressed if you use an approach to standard setting which considers and weighs both risks and benefits. The problems encountered in this type of approach are explained in this chapter, where we will consider the direct expenditures in cost-effectiveness analyses and the total benefits and losses in gain-loss balancing.

I. COST-EFFECTIVENESS

The cost of alternative actions and accompanying changes in risk are both parameters if one looks at risk acceptance in economic terms. Here we will discuss the use of costs of control, tax incentives (as alternatives to regulatory control), and ranges of benefits as economic factors in standard setting.

When our use of risk criteria alone is inadequate to establish acceptable risk levels, we may bring economics into consideration. One such approach is the *cost-effectiveness of risk reduction*.[52] Table 6 lists the methods one may use to measure when a cost-effectiveness level of risk is achieved. In this approach, the costs for various strategies are computed in dollar terms while risk reduction parameters are expressed in nondollar terms. Various actions to reduce risk may be ordered on the basis of the ratio of the magnitude of risk reduction to that of the cost of achieving this reduction. These individual actions are plotted in terms of this ratio as shown in Figure 3 and a curve plotted through them. Alternatives whose slope is increasing from left to right are eliminated and the resultant points form efficient cost-effectiveness points. When smoothed, the resultant curve is concave upward (Figure 3). Since the costs and risk represent the changes in going from one ordered efficient alternative to the next, the marginal cost of risk reduction is measured here, as opposed to the incremental cost of risk reduction. The latter assesses reduction from the initial to each level independently.

In each case, choices must be mutually exclusive to assure true marginality in the analysis. For example, if one uses two sequential steps of controls, A and B, and B could only be implemented if A were first implemented, then the resulting two cases would be A or A + B, but never B alone.

There are at least two approaches to cost-effective analysis. In the first case, the goal of risk reduction is not known *a priori*; in the second case, it is explicitly known. For the first approach, the curve reveals that the problem of assigning risk has simply been transferred

Table 6
COST-EFFECTIVENESS OF RISK REDUCTION

1. Technology limits
   A. Best available technology
   B. "As low as reasonably achievable"
   C. Best practical technology
   D. Limits of measurement
2. Cost benefit analysis
3. Values of lives saved
   A. Human capital approach
   B. Implicit social evaluation
   C. Insurance premiums and court decisions
   D. Risk approach
4. Economic
   A. Global cost-effectiveness for planning where to spend effort
   B. Marginal analysis after choosing appropriate scales
   C. Limit analysis taking ranges of uncertainties into account
   D. Discounting for the future
5. Taxes and other incentives
   A. Taxation approaches
      1. Risk "tax"
      2. "Polluter pays"
   B. Involuntary vs. voluntary compliance

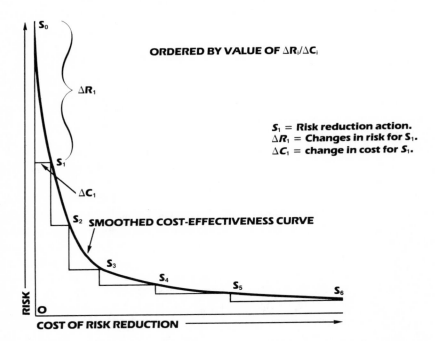

FIGURE 3.   Cost-effectiveness of risk reduction ordered relationship for discrete actions.

to a new parameter, the cost-effectiveness of risk reduction. A number of arbitrary conditions may be included (Figure 4), all involving external references. These may include "practicality of control", ability to measure, references to similar risks in society, and values of human life. Under all scenarios for cost-effectiveness of risk reduction, we need a referent before setting acceptable levels.

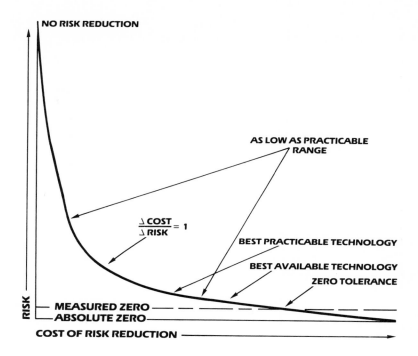

FIGURE 4. Some criteria for acceptance levels of cost-effectiveness of risk reduction.

A. Technology and *A Priori* Limits

Different types of technological risk references that have been used or considered are shown in Figure 4 and are described under the specific headings of technology limits. These are

- Best available technology
- As low as reasonably achievable
- Best practical technology
- Limits of measurement

If a goal is stated beforehand, such as the limit of dollars per risk reduced, the maximum dollars available, or the amount of risk to be reduced, alternate strategies may then be compared. Each of these strategies has a separate marginal cost-effectiveness curve as illustrated in Figure 5. Two alternative strategies, I and II, are shown, each with its ordered set of possible control options. Their cross-over point indicates an indifference level for either cost or risk reduction, but not both. Should there be only a fixed amount of money to spend, *a priori*, the horizontal fixed cost line shows two points (circles) which buy different amounts of risk reduction. Likewise, the vertical line indicating a fixed degree of risk reduction provides two alternate cost levels (crosses). The two squares illustrate the case where the ratio of the marginal cost and marginal risk reduction are equal to a specific value "a". The square indicates where the slope of lines, C/R, are equal to "a", in this case, a wide spread.

B. Cost-Benefit Analysis

In conducting a *cost-benefit analysis,* the expense of implementing a particular strategy is compared to ensuing benefits, in this case the risks avoided. In contrast to cost-effectiveness analysis, it is the usual practice to express benefits in dollar terms, thus ensuring that both

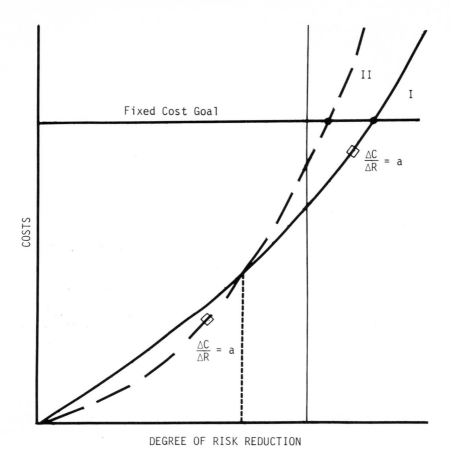

FIGURE 5.   Cost-effectiveness of alternate strategies for goals set *a priori*.

program costs and benefits are measured in the same units. Economists believe this is important when considering strategies with multiple outcomes, e.g., lives saved, injuries avoided, and structural damage averted, since they require some method for aggregating the various outcomes. Some reject this approach and see no apparent advantage in aggregation.

One must set a *value on life* which defines the benefit of risk reducing actions. If we set such a value, the cost-benefit balance is determined where marginal cost and benefit curves intersect, as illustrated in Figure 6. How does one determine the dollar value of a life saved?

C. Value of Lives Saved

Several approaches can be used to attempt to place a value on avoiding a statistically premature death. Four approaches considered in the literature are listed here.[13,53] These also involve value judgments, namely, an implicit social evaluation. Such evaluations often take place after a risk level has been set by other means, i.e., the value of lives saved is calculated from the decision. The four approaches are

1.   *Human capital approach* — The value of life is based on the premise that a man's worth to society depends on his productivity. As a productive unit, that worth is considered human capital.
2.   *Implicit societal evaluation* — Since society, through its political processes, does in fact make decisions on investment expenditures which occasionally increase or decrease

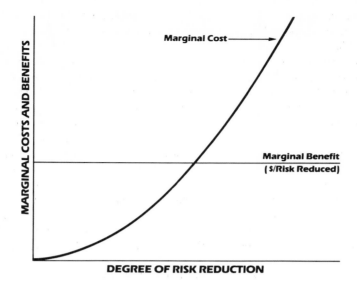

FIGURE 6.   Cost-benefit analysis.

the number of deaths, an implicit value of human life can be calculated. This does not require any direct calculation of the loss of potential earnings or spending. Instead, it approaches the problem from a social point of view by estimating the expenditure society actually makes to save a life. If, for example, an arrangement is made that will save an estimated five lives, at a cost of $100,000, then the implicit value of a life saved is $20,000.

3.   *Insurance premiums and court-decided compensation* — It is suggested that the amount of life insurance one is willing to purchase is related to the value one places on his or her life and the probability of being killed by some specific condition or activity.

4.   *The risk approach* — A more meaningful and often explicit measure is the amount of money society and the infrastructure are willing to pay to prevent a premature death. This can be observed by actually measuring what society pays for safety and anti-pollution measures. It is a derived measure.

When considering placing values on human life, recognize that there is a significant difference in how society regards a statistical premature death as opposed to an individually identifiable case.

D. Taxes and Other Economic Incentives

Establishing pollution taxes, pricing excess risks higher, and other economic incentives could be used to implement risk reductions and manage environmental quality.[54] Wilson has proposed a *risk tax* where the more risk involved in a product, the higher the tax.[18] Bower[55] analyzed incentives in a broader environmental context and pointed out the problems of too narrow an interpretation of impacts and risks suggesting that successful environmental quality management requires a simultaneous evaluation of three interrelated factors: physical measures, implementation incentives, and institutional arrangements. The recent experience where economic incentives were given in the form of higher water rates for excess water use shows the need for applying these incentives judiciously. Customers conserved so much water that utility revenues dropped, in turn endangering bond debt repayment schedules.

Pollution taxes for high risk projects may be publicly unacceptable, even if industry can pay, since they still allow some amount of environmental damage. This can be seen in the public response to uncontrolled discharges at Hooker Chemical facilities at Love Canal, N.Y. and the Life Sciences Kepone® intermediary manufacturing plant along the James River in Virginia.

A prospective use of economic incentives might involve setting an upper limit of discharge below acute toxicity levels. These incentives would then be used to protect public health by keeping actual discharges as low as practical below the upper level. They could be in the form of taxes, tax credits, reduction in administrative complexity, or direct payment of administrative costs to local governments for clean-up, surveillance, monitoring, and enforcement. Regardless of the form they take, they can be either linear with pollution increases or highly nonlinear to bias economic behavior toward low levels. For cases that can produce acute effects or are above existing standards, liability can be established and handled through existing judicial procedures. Thus, liability controls are indirect economic incentives.

Involuntary compliance is always more expensive than voluntary, but establishing the latter with either formal or informal regulations raises another question. How much enforcement is needed for a given degree of voluntary compliance? Here, one must balance the expenditure of "administrative costs" with the degree of desired goal attainment.

II. GAIN-LOSS BALANCING

So far, we have considered only direct expenditures for the purpose of reducing risks. There are some other benefits. The total spectrum of benefits are on the plus side of the ledger, as are gains from the activity. The costs, both tangible and intangible, and the risks are on the loss side of the ledger. Thus, losses include all the negative aspects. In any activity, both direct and indirect gains and losses must be included.

A *gain-loss balance* is one that, to the extent possible, encompasses all gains and losses. One approach is the sequential-feedback gain-loss analysis which involves four distinct steps:

1. Conducting a direct gain-loss analysis
2. Conducting an indirect gain-loss analysis
3. Determining the cost-effectiveness of risk reduction
4. Reconciling risk inequities

These steps are shown in Figure 7. First, we compare direct gains to direct losses. This is the classic trade-off analysis conducted before a new project or program is sponsored. The individual or institution undertaking the project receives project benefits, accepts project costs, and has full responsibility for making an initial, primarily economic, analysis. At this stage, voluntary risks, including investment, are taken to achieve specific results. If the direct gain-loss balance is negative, motivation for proceeding with the new enterprise disappears. Unless the balance is changed or new factors such as economic subsidization are introduced, the project will be discontinued. A balance in favor of gain, on the other hand, provides incentive for the program. Institutional factors such as legal constraints, tax incentives or disincentives, and public opinion are not always quantifiable at this stage. *Direct* gain-loss analysis is open-ended, because additional direct costs from subsequent steps may affect the gain-loss balance, and new factors that may change the balance must be evaluated as they occur. The sponsor, as the recipient of direct gains and losses, will continually review his position from the inception of the project through its completion, if only for economic reasons.

The second step in determining risk acceptance is to analyze the *indirect* gains and losses. Here we balance the indirect societal gains of a proposed activity against its indirect losses. (Note that risks themselves constitute but one aspect of the societal losses.) Such a balance must be made for local impact at a minimum and perhaps also for impact on the nation and

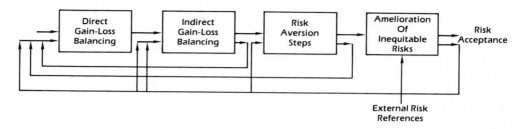

FIGURE 7. Risk evaluation process.

the world. Because of difficulty in quantifying these indirect effects, this type of analysis usually includes qualitative value judgments rather than numerical estimates. At the governmental level, a sponsoring agency is usually responsible for preparing the analysis.

The third step is determining the cost-effectiveness of risk reduction. Because society is generally risk aversive, the risk must be minimized to the extent feasible, even where indirect gain-loss balances favor gain. Thus, the expense of risk reduction must be computed for both direct and indirect gains and losses. The central question in this analysis is how to determine the point at which risk has been sufficiently reduced. Some other, more stable, reference is required to determine when risk reduction is cost-effective or when risk reaches an acceptable level. The development and use of such a reference, based on acceptable levels of *inequitably imposed risk*, is the heart of an effective methodology.

Thus, the fourth and most important step in evaluating risk acceptance is reconciliation of identified inequities. Even when an indirect gain-loss analysis is favorable, various inequities may be imposed on specific groups. They occur when those assuming risks do not receive benefits, or when risks are unevenly distributed among benefit recipients. In such circumstances, a risk that is otherwise acceptable because its gains outweigh its losses may become unacceptable because it is imposed inequitably. Inequities can be measured against references obtained from revealed, expressed, or implied preferences described previously.

In sum, the process for risk evaluation proposed here measures risk inequities against acceptable levels of risk in the form of risk references and other external criteria. This is done after the initial three steps in risk evaluation are performed. Each step requires readjustments in previous steps. This feedback is illustrated in Figure 7.

The difficulties of conducting an analysis of this type are immense. First, because there are many parameters involved, before aggregating to a single evaluation scale, one must establish relative weights and relationships. The choice of weights is a value judgment and can mask, or totally negate, the impact of key judgments which should be made at the highest decision-making level. Thus, retaining all important parameters as separate decision variables may be a better approach.

III. OTHER ECONOMIC ISSUES

Once a choice to use economic criteria is made, major issues of importance in implementing economic analysis may be addressed.* They are

1. Identifying relevant costs and benefits
2. Measuring costs and benefits
3. Selecting a discount rate to measure future costs and benefits
4. Determining the appropriate criterion for comparing costs and benefits

---

* Based upon an approach by Rachel Dardis.[56]

One can attempt to weigh benefits by first trying to estimate the monetary value of a life implicit in the decision to adopt various risk reduction strategies. In this procedure, the potential benefits (with the exception of lives saved) are assigned dollar values. The difference between total benefits and total costs decided by the number of lives saved provides an implicit value of a life. Strategies may then be compared with the preferred strategy possessing the lowest implicit value of a life.

The costs of any strategy are basically the required resources. They include direct expenditures for labor, raw materials, and other inputs. They are easier to measure than indirect costs such as the reduction in consumer choice when, for example, a particular product is no longer available to the consumer.

Because of reduced competition, some firms may be unable to comply with the proposed strategy and may exit the marketplace. The fact that a voluntary, rather than a mandatory, standard is in place may not prevent the situation from developing, since widespread adoption of a voluntary standard by most firms and product users could reduce the potential sales for a noncomplying firm.

Another indirect cost is the impact on innovation. The proposed strategy may either encourage or stimulate the adoption of new technology or the perpetuation of the status quo.

Identifying costs and benefits also lets one know whether they are borne by the same parties. If not, a particular strategy may be contested or become ineffective because of lack of support. Cost-benefit analysis can provide information concerning economic impact of a particular strategy on various groups and hence whether it may be acceptable.

The two remaining factors are the discount rate and the appropriate cost-benefit criterion. Some economists use the *present value of net benefits* (PVB) to evaluate different strategies. This is given by PVB = PVC where PVC is the present value of costs:

$$PVB = B_1/(1+i) + B_2/(1+i)^2 + \dots\dots\dots\dots\dots\dots\dots\dots\dots\dots\dots\dots\dots\dots B_m/(1+i)^m$$
$$PVC = C_1/(1+i) + C_2/(1+i)^2 + \dots\dots\dots\dots\dots\dots\dots\dots\dots\dots\dots\dots\dots C_n/(1+i)^n$$
$$B_t = \text{benefits accrued in year t, } t-1, 2\dots\dots\dots\dots\dots\dots\dots\dots\dots\dots\dots\dots\dots, m$$
$$C_t = \text{costs incurred in year t, } t = i, 2\dots\dots\dots\dots\dots\dots\dots\dots\dots\dots\dots\dots\dots, n$$

m = time period over which benefits accrue
n = time period over which costs accrue, and
i = discount rate.

As the above data indicate, the importance of the discount rate pertains to the manner in which benefits and costs accrue. If distributed equally over the life of the project, the discount rate we selected has little impact on the resulting evaluation. However, most benefits and costs occur at different points in time. The higher the discount rate, the lower the present value of future benefits and the greater the possibility that the project will be rejected.

Discounting for future risks has often been proposed. Can one discount human life in the same manner as economic investment and economic opportunity cost? There are many who argue that one cannot.[13,47,49] However, where it is accepted, the choice of which discount rate to use is critical. Changing a mere fraction of a percent and compounding the discount over long periods may make problems disappear on paper, but not in perception.

Global cost-effectiveness and *cost-benefit analyses* estimate the costs of all societal risks of a similar nature. Based upon results, priorities are set using most risk reduction per dollar expenditure as a criterion. For example, avoidance of high lead levels in some manufacturing operations may cost more per health effect avoided than would replacing old lead paint in houses. The difficulty lies not in the planning, but in the implementation. Institutional barriers are so well established that even reasonable consideration of such wide-scale effectiveness measures is not possible. In the U.S., for example, Occupational Safety and Health Agency (OSHA) is responsible for worker exposure, and the Department of Housing and Urban Development for housing regulations.

Still another type of assessment is the *limit analysis*. This analysis takes into account uncertainties in knowledge and judgment. Here we make worst and best case estimates in light of the limits of knowledge and the reasonableness for the purpose. We preserve the range of uncertainty until the final analysis. In some cases, uncertainty will have little effect on our decision since critical decision points lie outside the contested range. In other cases, the decision points lie within, and limit analysis does not provide a direct solution.

Part III

THE ANALYSIS

There is always something to upset
the most careful of human calculations.

Ihara Saikaku (1642—1693)
The Japanese Family Storehouse;
or, The Millionaires' Gospel, bk. II,2

We are now ready to assess the various methods of setting levels of risk for standard setting. However, before this is done, if you have not done so, we strongly advise you to read the introduction to this book. There we explain the reasons for selecting cadmium and phenol as "strawmen" and describe the type of toxic material each represents. The broad conclusions of our study presented there will assist in adapting the principles and steps described on the following pages to specific problems or to an overall understanding of standard setting.

As a first step, one should determine the procedural and intellectual framework to follow in conducting a particular analysis. In other words, what models should be used. Models selected may, of course, be modified later. We have developed models which, while directed at our particular purposes, should show the application and limitations of such models in general. Prior to or roughly concurrent with model development, preliminary data should be gathered and information distilled in certain key areas. The type of model chosen may depend on what is learned from those data.

At this point, one should determine the industrial pollution sources via the investigated medium; in our case, water. This ensures early determination of contributing sources, necessary before constructing models. The impact of other water-borne sources, both man-made and natural must also be considered. We can then compare the impact of water-borne residues from industrial activity to that in other media and from other sources.

As in the determinations of levels for actual standards, our assessment of methods requires gathering and assessing the physical properties and environmental fate potentials of the selected water pollutants. This is necessary before estimating human and wildlife exposure. Given all this information, we may then construct models for later use in analyzing risk from the sources and pathways we are considering. These models form the framework of the subsequent analysis.

Chapter 7

THE NATURE AND POLLUTION SOURCES OF CADMIUM

To waste, to destroy, our natural resources, to skin
and exhaust the land instead of using it so as
to increase its usefulness, will result in
undermining in the days of our children
the very prosperity which we ought by right to hand
down to them amplified and developed.

Theodore Roosevelt
Message to Congress
December 3, 1907

The persistence and resultant accumulation of cadmium makes it of concern to those who weigh the impact of the pollution of today on future generations. This chapter provides an understanding of this metal and illustrates the types of information acquired in a risk assessment.

I. SOURCES OF EXPOSURE

To determine the importance of *exposure sources*, we must consider both the responsibile entity and the route, or *pathway*, by which the substance is transmitted to a receptor. Both will be discussed in this chapter.

A. Pathways

Although one need consider only exposure via the medium for which a standard applies, water contamination may result from many sources and these must all be examined. The major routes by which man may be exposed to cadmium-contaminated water are
1.   Contaminated irrigation water (exposure via food and runoff to streams)
2.   Watering of crops and stock (bioaccumulation and runoff), as well as
3.   Direct and indirect (via POTWs) contamination of water and accidents
Direct contamination of water leading to residues in drinking water is the most significant source of human exposure as well as the one which can most readily be controlled through industry standards. We will focus on these pathways. Spillage appears to be a rare event.
Contamination by end-use products is of peripheral concern and we will briefly address it later. We will not consider differential effects from the various types of cadmium.
Once one determines the exposure routes, information is gathered on the human and wildlife populations at risk. This will be needed in constructing models. In our case, we used statistics on waterflow in U.S. rivers, populations in average-sized municipalities, and the percentage of water cities obtain from ground and from surface supplies in determining the human populations at risk. This information is given under the discussion of models.

B. Industrial Sources

The next step is to identify the major industrial sources of cadmium water pollution, a process providing amounts and pathways of effluents, values of technologies, and the feasibility of using substitutes. The latter two types of information are used in determining economic need and impact.
Cadmium exposure, as we have described, results from both its production and consumption (i.e., commercial use). The production sources, mining and smeltering, are the primary polluters of water and air, respectively. Consumption contributes primarily to land/landfill pollution. The inadvertent presence of cadmium in the secondary metals industry,

---

OTHER RELEVANT ASPECTS OF CADMIUM POLLUTION

- Water contamination may occur from decomposition of cadmium products (for example, batteries).
- The electroplating industry discharges a portion of their effluent directly into municipal water treatment facilities.
- Airborne cadmium particles and cadmium deposited on land may significantly add to water concentration (discussed under "Environmental Fate").
- Phosphate fertilizers are a potential source of water contamination despite the fact that most cadmium from this source is taken up by plants.
- Cadmium used by the pigment industry, unlike that used in batteries or by the electroplating industry, is in an insoluble form with less potential exposure to man.

---

coal-fired power plants, and phosphate fertilizers may also pollute land, air, and water. Incineration of wastes and disposal of effluents in sludge release cadmium to air and land, respectively. Residues are distributed during the waste disposal of cadmium products and effluents. The variety of these sources makes it difficult to pinpoint major polluters and pathways to man.

The major primary and inadvertant emitting industries selected are

1. Mining and smelting of zinc and cadmium
2. Electroplating
3. Pigments
4. Plastics
5. Ni-Cd batteries
6. Alloys
7. Phosphate fertilizers
8. Secondary metals

We must also determine the number of plants or polluting entities and the water emissions for later use in a model.

II. ENVIRONMENTAL FATE

Cadmium occurs in the environment as a result of both natural and man's processes. It occurs naturally mainly as a component of minerals in the crust of the Earth with an average concentration of 0.18 ppm. The pure metal does not exist in nature, but minerals containing cadmium are always associated with zinc ores. These include zinc sulfides, zinc oxides, zinc silicates, and polymetallic zinc ores. The cadmium content is related to the content of zinc, with zinc to cadmium ratios ranging from 100:1 to 1000:1.[44]

Cadmium is also present in relatively low concentrations in igneous rocks. Uncontaminated soil usually contains less than 1 ppm cadmium[44] and levels less than 1 ppb occur in non-polluted sea and fresh waters.[57] Exceedingly low cadmium concentrations occur in rural atmospheres.

Metallic cadmium is prepared commercially as a by-product of primary metal industries, principally the zinc industry. Cadmium is found not only in zinc ore, but in lead, copper, and other ores containing zinc minerals.[44]

The major cadmium input to coastal waters is leachate from mine tailings; disposal of waste streams from hydrometallurgical plants, electroplating processes and plant manufacturing cadmium-containing pigments, cadmium-stabilized plastics, and nickel-cadmium bat-

CHEMISTRY OF CADMIUM

Cadmium (Cd), a chemical element of atomic number 48 and atomic weight 112.4, occurs in 8 naturally occurring isotopes of varying abundance, [1-6]Dc (1.22%), [108]Cd(0.88%), [110]Cd (12.39%), [11]Cd(12.75%), [112]Cd(24.07%), [113]Cd(12.26%), [114]Cd(28.86%), and [116]Cd(7.58%). The electron configuration of the atom is 2, 8, 18, 18, 2. The only valence of importance is the +2 state. This Cd^{2+} ion has an outer shell of 18 electrons and an ionic radius (unhydrated) of 0.97 Å.[44]

Cadmium is a bluish-white-silver metal with a melting point of 321°C, boiling point of 765°C, heat of fusion of 13.6 cal/g or 1.53 kcal/mole, heat of vaporization of 213 cal/g or 23.9 kcal/mole, and Brinell hardness of 21. Its vapor pressure is 1.4mm at 400°C and 16mm at 500°C. Cadmium fumes are released in thermal treatments such as ore roastings, brazing, re-smelting of steel scrap, and incineration of cadmium-containing refuse. The resulting vapor is reactive and in air quickly forms finely divided cadmium oxide.[58]

teries; disposal of sewage effluents and sludge; surface run-off in urban areas; and deposition of atmospheric soluble cadmium sulfate and chloride compounds.[57]

Although water may not have high concentrations, the sediment may be highly contaminated, particularly at a neutral or alkaline pH.[59-61] To avoid errors when determining the degree of contamination in water, the amounts of cadmium in suspended particles and in sediments must both be measured.

Settling into bottom sediment attenuates the amount of available cadmium in the water. STORET (STOrage and RETrieval) data compiled by EPA are a good source of the amounts in solution and in sediment and were used here in designing a model for cadmium.[62]

It is important to know for this analysis that municipal water treatment plants, other than those employing sand filtration, cannot remove cadmium. On the average, 80 to 90% passes through and, in some instances, damages the system. In facilities employing sand filtration, less than 20% of the cadmium passes through. In all cases, disposal of sludge containing cadmium may present a problem.

Less than 3% of the total cadmium in runoff is from a pathway other than direct ground contamination.[63] It can, therefore, be concluded that for most of the U.S., water contamination via ground primarily results from direct ground, not air contamination. Areas around zinc smelter plants and incinerators would be expected to have more cadmium in circulation from air to ground to water.

ADDITIONAL SOURCES OF WATER
CONTAMINATION FROM CADMIUM

Water contamination cannot be solely attributed to direct disposal since air and ground contamination may also contribute. Most municipal water supplies are derived from surface water susceptible to deposition of airborne contaminants.[64]

Water pollution depends primarily on local factors such as upstream sources, subsurface contaminants, and direct discharges. Water is contaminated with cadmium (1) directly, (2) from the air, (3) from the ground, and (4) from air indirectly via ground sources. Direct or water borne effluents constitute only 1 to 2% of total cadmium emissions. Air emissions account for 16 to 17% with the remainder destined for land.[60,65]

Ground contamination with cadmium is a result of the dumping of wastes, eventual disposal of products, air deposition, and use of phosphate fertilizers. Waste disposal is a minor contamination source, since wastes usually pass through sewage treatment or incineration/open burning before deposition on land. Approximately 80% of the cadmium end-use products are eventually disposed in landfills. In fact, 95 to 96% of the cadmium used is eventually deposited on land, whether it be rural, urban, or fill.[66] The addition of phosphate fertilizers can increase cadmium soil concentrations by a factor of 10.[60]

As a result of man's and nature's activities, cadmium is released from various compounds and gradually accumulates in the environment, including the tissue of aquatic organisms.

ORGANIC REACTIONS OF CADMIUM

A bacterium, *Pseudomonas* sp., isolated from sediment from the Chesapeake Bay was able to methylate cadmium.[67] Based on acute toxicities of other methylated heavy metals (e.g., methyl mercury), the potential detrimental effects of methylated cadmium on the biosphere cannot be taken lightly. When setting a standard for cadmium, an attempt should be made to quantify this effect as far as possible.

Chapter 8

EXPOSURE TO CADMIUM

Water, water everywhere,
Nor any drop to drink.

Samuel Taylor Coleridge
The Ancient Mariner
(1798) pt. II, st. 9

Obviously, one cannot avoid exposure to substances in water unless a cleaner source of water is available.

An analysis of risk should consider risk from the investigated pollutant in both ambient and effluent water, and the degree of additional risk exposure to water contamination adds to that via other pathways. An exposure analysis must address, in some manner, special populations with increased risk because of either greater exposure or sensitivity than the general population. These factors are explored in this chapter.

I. AMBIENT LEVELS

The term *ambient* is generally applied to background, sometimes naturally occurring, amounts of a substance. Ambient amounts are often subtracted from total monitored amounts to determine how much pollution may be attributed to a particular cause or source. In this analysis, we are interested in the additional cadmium contributed by industrial sources. Ambient will here refer to the existing levels of pollutant.

A. Drinking Water

We have extrapolated ambient levels of cadmium in drinking water for some municipalities to the general population, estimating an average daily consumption of 2 ℓ of water. We used the draft of a survey of organic, metal, and other pollutants in selected water supplies in EPA Region V (EPA, 1975) in obtaining these levels.[68]

Region V includes the states of Illinois, Indiana, Ohio, Michigan, and Wisconsin. Although selection of cities for sampling water supplies was not based on population, Table 7, taken from the published survey, indicates that a complete range of population sizes was represented. Of the total Region V population of approximately 44 million, 34% or 15 million people were included in this survey.

All samples were analyzed for cadmium. The range in finished water was approximately 0.0002 to 0.0004 mg/ℓ. This narrow range is somewhat surprising because of wide differences in number and types of local industries, population size, and wide variability documented by other authors.

Although the survey was conducted during winter months when certain industrial activities were at a minimum, the 0.0004 mg/ℓ level is adopted here as a worst case ambient level.

B. Raw Water

We derived estimates of the current residues in U.S. bodies of water from analyzed samples recorded in the Environmental Protection Agency STORET storage and retrieval system.

The mean total suspended and dissolved cadmium in U.S. raw water* samples of 5.1

---

* Water in nature, unprocessed by man.

Table 7
POPULATION OF CITIES IN EPA
REGION V WHERE DRINKING
WATER SUPPLIES WERE
SAMPLED IN 1975

Population class in thousands	% of Cities sampled	% of Population included in study
1000—4000	2.4	48.8
500—1000	4.8	18.5
100—500	16.9	20.5
50—100	19.3	6.7
0—50	56.6	5.5

Table 8
CADMIUM LEVELS IN RAW WATER
SAMPLES, STORET, 10/62 TO 2/80

Cadmium	Collected	Mean	Maximum	Minimum
Dissolved	10/62 to 11/79	3.97 μg/ℓ	500.00 μg/ℓ	0.00 μg/ℓ
Suspended	5/69 to 10/79	5.78 μg/ℓ	320.00 μg/ℓ	0.00 μg/ℓ
Total	7/68 to 2/80	5.1 μg/ℓ		

μg/ℓ, as measured by STORET, is used as the ambient raw water level.[62] These are shown in Table 8.

II. OTHER PATHWAYS

We must now determine residue levels of cadmium in some of the more important human exposure pathways. Appendix III-B lists potentially more sensitive members of the population and both likely and "worst case" exposure via a number of pathways. Remember that we are obtaining hypothetical numbers for purposes of this study.

There are considerable data on residues in foodstuffs. Estimates of the average dietary intake for a 15- to 20-year-old male in the U.S. vary widely. A high estimate is 100 μg/day. However, up to 15% of the population eating higher proportions of certain foods may be exposed to as much as 300 μg/day.[69] Meats, chicken, dairy products, and seafood may have especially high residues. Samples of oysters have been found to contain 3.66 μg/g, by far the highest level for any food sampled.[70] Several vegetables, especially spinach, may also have relatively high levels but do not approach levels in the former group.

Exposure, via inhalation, may be substantial. A 1972 air analysis in East Helena, Mont. showed a mean cadmium content of 0.29 μg/m³ with a high of 0.69 μg/m³. This compares to a range of 0.002 to 0.1 μg/m³ in air from 11 rural areas, presumably remote from cadmium-emitting factories.[60] Smoking of tobacco presents an additional inhalation risk; 20 cigarettes, 1 package, contain a mean cadmium content of 16 μg. Between 2 and 4 μg are inhaled. Inhalation is also the primary route of occupational exposure.[71]

Dermal exposure may occur by swimming or other recreational uses of polluted waters and is, of course, generally dependent on the stirring up of sediment containing cadmium. Because cadmium is persistent and may move into areas relatively free of industrial pollutants, this presents a potential exposure route.

Chapter 9

THE NATURE AND POLLUTION SOURCES OF PHENOL

I counted two-and-seventy stenches,
All well defined, and several stinks.

Samuel Taylor Coleridge
Cologne (1828)

Things are generally not this bad in U.S. waterways, but in phenol we have a compound whose metabolites produce disagreeable odors in water. Unlike cadmium, phenol is organic, a compound not an element. It is also nonpersistent (a relative term, of course) in water, readily forming various phenolic compounds, some of possible risk to humans. The question of how to determine exposure and risk from such a compound, and then how to regulate its environmental contamination, has been a problem to regulators.

I. SOURCES OF EXPOSURE

As for cadmium, we will consider both the pathway and cause of exposure to phenol.

A. Pathways

The exposure routes by which phenol-contaminated water may pose an exposure risk to man are the same as those for cadmium. They are

- Use of contaminated irrigation water
- Watering of stock
- Direct contamination of bodies of water
- Indirect contamination of bodies of water (via POTWs)
- Contamination of water supplies
- Accident

Again, we are most concerned with direct water contamination from industrial wastewater and whether it reaches humans in drinking water.

Another means by which contaminated water may pose an exposure risk to man is by accidental spillage. This appears to be a rare event with regard to phenol. In the history of phenolic use, only two major accidents are known to have occurred. The first resulted from the failure of a dike to hold industrial waste waters in Luxembourg in the early 1960s.[72] The second occurred in the small community of Lake Beulah, Wisc. in 1974, when a major train derailment dumped 37,900 lb of 100% phenol. Saturation of the soil caused pollution of well water, which, despite dilution efforts, resulted in detectable phenol levels for 2 1/2 years.[73,74]

As with cadmium, model construction requires information on the population which could be exposed via the pathways selected.

B. Industrial Sources

The phenol emitting industries we will address are

- Phenol manufacturing
- Coke (with and without "dephenolization"*)

---

* A 2-stage process to remove phenol.

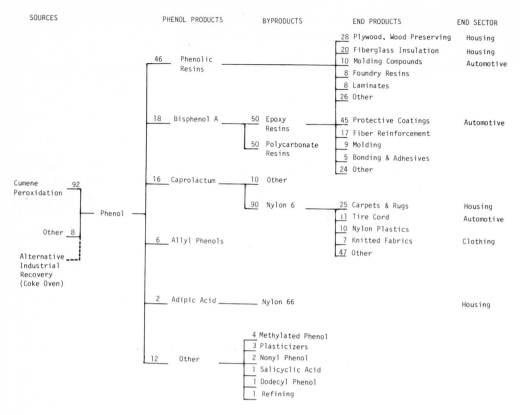

FIGURE 8.   Manufacturing flow and major end uses of phenol and phenolic compounds.

- Petrochemical
- Plywood
- Phenolic resins
- Oil refineries
- Blast furnaces

Phenol appears to have an almost infinite variety of end uses. Figure 8 charts the production and uses of phenol and the ultimate products. The two major users are the housing and automotive industries. Although some phenol is utilized to produce pesticides, pharmaceuticals, and other products, the primary products are resins and resin bases. In the resin industry, phenol is a versatile feedstock because of its distinct physical properties. Phenol application is typically product-specific and in most cases, may not be readily replaced.

As a common water-soluble organic chemical, phenol enters the environment through both economic and noneconomic (indirect) pathways. For example, it is emitted as a waste product by coke ovens, blast furnaces, and oil refineries although not directly utilized in the manufacturing process. When high concentrations exist, phenol is often recoverable. In fact, an indirect phenol source is the recovery of phenolic effluent from the waste water of coke ovens. This process may be an economically feasible means of reducing wastes, but does not currently provide a competitive production source of phenol.[75,76]

The major source of phenol in waterways is from this production of coke which is primarily used in blast furnaces.[75] Data on the amount of phenol emitted from coke ovens and blast furnaces are, therefore, redundant to some extent when considering human exposure. In addition to these industries, oil refineries emit a multitude of organic compounds including phenol.[77]

CHEMISTRY OF PHENOL

Although the term "phenol" is often used to describe compounds containing one or more hydroxyl groups attached to an aromatic ring, we will use it in specific reference to monohydroxylbenzene, or carbolic acid. Phenol has an empirical formula of C_6H_6O, a molecular weight of 94.11, and is composed of 76.57% carbon, 17.00% oxygen, and 6.43% hydrogen. At 25°C it is a clear, colorless, hygroscopic, deliquescent, crystalline solid; however, if impurities are present, it becomes pink.[85,78]

Phenol exhibits weakly acidic properties with a pK_a (acid dissociation constant) at 9.9. It melts at 40 to 41°C, boils at 181.75°C, and has a vapor pressure, in mm of mercury, of 0.35 at 25°C. At 25°C in solid form, its specific gravity is 1.071, in liquid form, 1.049. Phenol has a high solubility in water of 6.2 g in 100 mℓ at 16°C. The amount dissolved increases with temperature. Phenol is also soluble in organic solvents such as benzene, alcohol, chloroform, ether, glycerol, carbon disulfide, and petrolactum. The octanol-water partition coefficient is 31.[85,78]

The petrochemical industry produces raw phenol primarily by the perioxidation of cumene, a second-generation petrochemical product derived from benzene and propylene.[78,79]

Plastic resins, such as the epoxy and polycarbonate resins and the general class of phenolic resins, are the major products derived from phenol. Nylon feedstocks, such as caprolactum and adipic acid, are also produced from phenol. These have extensive use throughout the economy, an important fact when analyzing the impact of variations in phenol price and availability on user industries.[79]

Phenol, a benzene-derived compound, differs from cadmium in that it is environmentally degradable; cadmium is a metallic element with no significant nuclear decay (it is not radioactive). As a weak acid, phenol has a reactive affinity with halogens and serves as a foodstuff for certain microorganisms. It is volatile in chemical and biological environments such as waterways, not accumulating in ecological and hydrological cycles. Because of this tenuous and brief existence at large concentrations, it is difficult to ascertain the impact on man and the environment of phenol pollution through waterways.

As with cadmium, we must determine the number of polluting entities and the emission via the pathways investigated at this point.

II. ENVIRONMENTAL FATE

Phenolic compounds are produced naturally by plant and animal decomposition. There are few natural sources of phenol per se. It does occur in the needles of pine cones *(Pinus sylvestris)*, the essential oil of the leaves of tobacco, *(Nicotiana tobacum)*, of currant *(Ribes nigrum)*, and of the herb *(Ruta montana)*, and in lichen, *(Evernia prunastri)*.[80]

Commonly used analytical methods are not specific for phenol.[81] In fact, the classic detection analysis is both nonspecific and insensitive to some common phenolics.[82]

Phenol may enter waterways through sewage, manure, soil, vegetation, or industrial wastes. The last is by far the most serious contamination source. Municipal sewage adds to this contamination when industrial wastes discharged to municipal treatment facilities are high in phenolic content.

Phenolics are one of the most common classes of toxic organic compounds found in waste waters.[83] Phenol is the most prevalent chemical of this class being present in domestic waste water at concentrations of from 0.1 to 1.0 mg/ℓ.[84] Effluent wastes containing phenol are generated by coke ovens, oil refineries, petrochemicals, aircraft maintenance, fiberglass manufacture, and the plastics industry.

Microorganisms degrade phenol in waterways at rates between 2 to 5 mg/ℓ/day.[75] This information is of particular importance in constructing our model.

EPA CRITERIA FOR PHENOL

The most recent EPA criterion for phenol is set at 3.5 mg/ℓ to prevent most toxic effects to saltwater and freshwater aquatic species.[86] The 1975 criterion was 1 μg/ℓ for both surface waters and domestic water supplies where chlorination is used.[87] For comparison, that agency estimated that a level of 0.3 mg/ℓ would be necessary to control undesirable taste and odor of ambient water. Phenol also reacts with chlorine in standard drinking water treatment to form even more odorific compounds.

The EPA has observed that there is no demonstrative relationship between organoleptic and human health effects.

Because phenol per se has few natural sources, it is not a normal constituent of drinking water and is present only at relatively low levels in most surface waters. In the absence of known sources, it does not usually occur in drinking water supplies above the analytical detection limit of 0.001 mg/ℓ.[73] Most studies of natural aquatic ecosystems analyze concentration of total phenolic compounds rather than phenol alone.

Chapter 10

EXPOSURE TO PHENOL

For all we take we must pay,
but the price is cruel high.

Rudyard Kipling
The Courting of Dinah Shadd (1890)

Part of the price an industrial society must pay may be some measure of exposure to hazardous chemicals. In this chapter, we estimate human exposure to phenol via water.

I. AMBIENT LEVELS

The differing natures of cadmium and phenol result in quite different ambient levels.

A. Drinking Water

Phenol drinking water levels were obtained from the 1976 EPA-conducted National Organic Monitoring Survey which compiled concentration data on 32 organic compounds in the finished water supplies of 113 communities. Phenolic compounds sampled were 2,4 dichlorophenol and pentachlorophenol which are among the chlorinated phenols common in chlorine-treated water. The minimum detectable concentration for both compounds was 0.10 mg/ℓ.[88]

We derived population figures, based on 1977 data, by assuming the 113 samples characteristic of the overall U.S. metropolitan population. Figure 9 presents that extrapolation.

Results indicate that less than 8 million people, or 5% of those living in metropolitan areas, are exposed to a level exceeding the 1975 EPA limit of 1 mg/ℓ. That level was based not on physiological consequences, but on odor and smell criteria. The maximum levels were 1.3 to 1.4 μg/ℓ recorded in Phoenix, Ariz. and Tampa, Fla.

B. Raw Water

STORET data accumulated between September 1977 and March 1978 are used as the ambient raw water level. The mean level of samples analyzed is 11 μg/ℓ.[89] Data appear in Table 9.

The total phenol level included suspended particles as well as dissolved phenols. STORET records levels of chlorophenols and other phenolic compounds separately from those of phenol.

II. OTHER PATHWAYS

Data on exposure to phenol, unlike that for cadmium, are sparse. We have, therefore, conducted a sample exposure analysis for phenol. This material appears in Appendix III-A. Figures cited in this chapter are derived from that analysis.

The Market Basket Survey of foodstuffs does not include detection of phenol or phenolic compounds. We have assumed dietary levels are from ingestion of red meat and finfish. It is unlikely, given the characteristics of these chemicals, that phenolic contamination in foodstuffs other than animal tissue is either widespread or high. Since animals have phenol present as a normal metabolic product, one would expect residues in meat samples. We have used levels in two meat samples[90] to derive an average 12 mg/kg for meat and meat products in the diet, recognizing that this is hardly a representative sample.

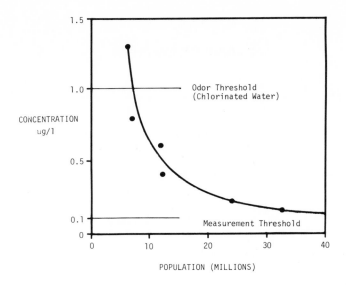

FIGURE 9.   Exposure to phenol via community water supplies.

Table 9
PHENOL LEVELS IN RAW WATER
SAMPLES, STORET, 9/77 TO 3/78

Phenol	Mean	Maximum	Minimum
Dissolved	0.10 μg/ℓ	0.10 μg/ℓ	0.10 μg/ℓ
Total	11.00 μg/ℓ	28.00 μg/ℓ	0.00 μg/ℓ

Fish have varying uptake depending upon the species and water conditions. The sum of derived dietary exposure used is 0.1 mg/day for finfish and 2.7 mg/day for meat and meat products. Consumption of finfish, of course, represents an exposure resulting from water contamination.

Dermal exposure results from use of medical preparations containing phenol. (Appendix III-A lists some of these products with phenol content). A worst case assumes multiple use of these preparations. Occupational exposure to phenol is also primarily dermal.

Except perhaps for inhalation in enclosed industrial settings, relatively rapid decomposition in the presence of oxygen makes inhalation of this pollutant of negligible consequence to the general population.

Chapter 11

USE OF MODELS

(Concerning the Six Principals of Painting)
The first is, that through a vitalizing
spirit, a painting should possess the
movement of life.
The second is, that by means of the
brush, the structural basis should be
established.
The third is, that the representation
should so conform with the objects as
to give their likeness.
The fourth is, that the coloring
should be applied according to their
characteristics.
The fifth is, that through organization,
place and position should be determined.
The sixth is, that by copying the ancient
models should be perpetuated.

Hsieh Ho, A.D. 500
Notes Concerning the Classification
of Old Paintings from The Spirit of the Brush,
translated by Shio Sakanishi (Widsom of the East Series, 1957)

The painters of today mix their brushes
and ink with dust and dirt, and their
colors with mud, and in vain smear
the silk. How can this be called
painting?

Chang Yen-Yuan, c. 850
Discussion of the Six Principles of Painting

Painting — a far cry from modeling? Not really. Painting, like modeling, attempts to capture reality, its essence or some aspect of its being. As with painting, modeling follows some guidelines or principles (the comments of Chang Yen-Yuan need no elaboration). Both may be viewed subjectively as to whether they represent what is intended.

Models provide a framework for our analysis of risk levels in standard setting. One must have a method for determining the amount of a pollutant to which various segments of the population are exposed.

Remember that we are concerned with procedures for setting levels for standards to reduce water-borne risk from two very different chemicals. We have pursued the major aspects, but not every aspect, of the problem. Instead we will call your attention to additional areas of consideration.

Two models are presented. The first, the "river" model ties water pollution by major industrial polluters to the amount taken in by municipalities downstream. An exposure model estimates the number of people exposed and the degree of exposure. There are several smaller models which provide input to these major models. The models disussed here are termed "effects/impact" models. The use of these and other types of models is discussed in Appendix IV-A.

I. RISK ESTIMATION MODEL

To estimate the present level of risk from effluents, one may design a model tracing effluent concentrations and volumes at discharge points through the environment to humans. To reiterate, the model developed here is for evaluating standard methods of setting levels for standards. Although not precise for estimating absolute risks, it should be adaptable to establishing relative human risk for water pollutants. It may be modified for other pathways and expanded to include other species. Because of the many plausible but unvalidated assumptions in the model, these particular risk estimates may bear little resemblance to reality. For this reason, these estimates should be used only in this particular evaluation.

There are three models, or sub-models which comprise the risk estimation model. Inputs to the model are the effluent concentration and volume from each industry using an average capacity and concentration level at a river discharge point. A *river model* accounts for dispersion and attenuation of the effluent in both low-flow and average-size rivers. Resulting concentrations are used as input to another model, the *exposure model*. Since more than one industry may contribute to the exposure of a given population, an *overlap model* has been developed to account for this problem in a limited manner.

II. RIVER MODEL

The purpose of this model is to link industrial pollution with the point of municipal uptake in a typical waterway. The important variable is the ultimate downstream concentration at point of uptake. A pollutant emitted into a waterway at a given flowrate and concentration is reduced by two factors.

1. Mixing in the waterway
2. Further attentuation due to physical or chemical factors

There are essentially three parameters influencing downstream concentrations.

1. Type of industry
2. Characteristics of the waterway
3. Type of pollutant

Different industries generally have different water usage demands, as do individual plants with different production capacities. The pollutant flowrate and concentration injected into the waterway therefore depends on industry type.

Waterways also differ in average flowrates and seasonal variations. The flow velocity and the amount of sediment carried will, of course, affect mixing. Flowrates are variable, changing with location along a given water — the mouth and source being the extremes.

The pollutant itself may be soluble or may form an emulsion which inhibits mixing. Its chemical and physical characteristics also influence the downstream concentration. Cadmium, for example, tends to settle and accumulate in deposited sediment because of its large molecular weight and frequent discharge as an insoluble sludge while phenol serves as a food supply for certain microorganisms.

Preparation of a river model consists of four steps.

1. Identifying the major industries that emit the pollutant (this has been done in Chapters 7 and 9)
2. Determining the waterway characteristics
3. Computing the concentration dilution
4. Determining further attenuation

Table 10
FLOW DATA FOR CHARACTERISTIC RIVERS

Category	River	Flow	Low flow	High flow
Largest	Mississippi	214,000	100,000	380,000
waterway	Ohio	120,000	33,000	250,000
	Missouri	23,000	11,000	38,000
Smallest	Humbolt (Nevada)	130	16	320
waterway	Colorado (Texas)	1,800	1,000	3,000
	Grand (Michigan)	2,200	940	4,900
Average	69 rivers	17,700		
waterway	Cumberland (Kentucky)	17,800	4,300	40,000
	Wabash (Illinois)	17,000	4,400	31,500
	Tombigbee (Alabama)	17,000	3,700	40,000

Note: All units in MGD-million gallons per day.

(After van der Leeden[92])

Calculations are based on the untreated incremental addition from a representative plant in each industry. These plants are assumed to be discharging into a representative waterway upstream from a representative municipal locality, a Standard Metropolitan Statistical Area (SMSA).[91] We will determine both average effluent concentration and average flowrate for each industry, with the understanding that there is substantial variation on each dimension.

A. Data Base

We have developed surrogate waterways from statistics on waterflow in 69 principle rivers throughout the U.S.,[92] using three waterway categories: largest, smallest, and average; and three flowrates: normal, seasonal low, and high. Results are listed in Table 10. Several flowrates at different river sampling stations are used in determining overall average for the largest waterways. For the smallest waterway, there is wide variation in flowrate. That river, the Humbolt, is by far the worst case situation in terms of low flow. In fact, the two other rivers in this category may be more representative of a typical small river used as a wastewater discharge sink. The average waterway category was obtained by simply averaging the mean flowrates for all rivers with the exception of several Alaskan rivers. The latter were excluded because they were unlikely wastewater sinks and their large flowrates would bias results. The Cumberland River was found to be representative of the average waterway and is used as the surrogate waterway epitomizing U.S. rivers.

Of the categories in the table, two will be used: the average waterway, the Cumberland, at normal flow and the smallest waterway, the Humbolt, at normal flow. The largest waterway category exhibits such large flowrates that resulting downstream concentrations are beyond the measurement threshold of most pollutants. The average and smallest categories provide more significant information. The worst case situation is best exemplified by the smallest waterway at normal flow because the low flow value for these waterways is extreme.

B. Calculations

Dilution of effluent in the smallest waterway is less than that of the average waterway by a factor of about 100. The actual dilution ratio from plant discharge to downstream is computed using the following equation:

$$D = \frac{F \text{ water}}{F \text{ plant}} + 1$$

where the ratio of the waterway flowrate (F water) to the average plant flowrate (F plant) is computed for each industry. We then divide the average pollutant concentration for an individual plant by the dilution ratio to obtain downstream concentration after complete mixing. The problem of diffusion is accounted for by assuming as typical a 1-day retention period before municipal uptake. This should provide ample time for complete mixing.

In some industries, there are more plants than municipal localities, as represented by SMSAs. To prevent overlapping, we assume that in these industries more than one plant is discharging into the local waterways. Since there are only 279 SMSAs, this serves as the maximum number of "exposure sets" for the U.S. urban population. Industries with more than 279 plants require that we correct dilution ratios by the following factor:

$$\text{F plant} = \frac{N}{279} \text{ (F plant)}$$

where N represents the number of plants in the industry. The factor represents the expected average number of plants per exposure set and the corrected flowrate and total discharge rate for each exposure set.

The downstream concentration due to mixing is simply discharge concentration divided by dilution ratio, D. Attenuation or reduction, in the amount present from nonmixing factors is included to determine concentration before municipal uptake.

C. Attenutation

Biological degradation by microorganisms is a major factor in phenol *attenuation*. This degradation rate in streams ranges from 2 to 5 mg/ℓ/day.[75] Phenol-oxidizing microorganisms are not dominant in aquatic environments with low phenol concentrations and many cannot survive above 150 ppm phenol.[93] We may assume that a partially diluted concentration of about 100 ppm has a half-life of 10 days, based on the above degradation rate. Thus, the 1-day waterway retention attenuates concentrations about 5%. An additional 1–day retention period is assumed for water being dispersed through the municipal water supply system. Thus, 5% attenuation occurs in the waterway and 10% along the entire pathway from plant to distribution point. The attenuation model assumes phenol biological degradation can be expressed in terms of a fractional reduction; in other words, the phenol-consuming mircoorganic population is essentially proportional to phenol concentration, which serves as a food supply.

Cadmium attenuation, on the other hand, results from the settling of dissolved cadmium into the floor sediment of waterways. We assumed that virtually all residues deposited in sediment are the result of waste discharge into waterways. STORET data on average ambient cadmium concentrations indicated that for a 5.10 ppb total concentration, 3.97 ppb exists as a dissolved solute. Thus, we can assume approximately 24% reduction is due to settling. Numerous studies indicate that a good portion settles out within one day. We further assume that 50% settles in the 1-day retention period before municipal uptake, yielding an attenuation factor of about 12%.

III. EXPOSURE MODEL

The model for determining exposure level and numbers exposed uses as input the attenuated concentrations for each industry category from the river model. The main route of exposure is surface drinking water at an average adult rate of 2 ℓ/day.[94] An SMSA of typical size receiving an attenuated concentration from an upstream industrial source is the population base. However, many water supplies are from ground water for which no pathway exists for either cadmium or phenol from industrial effluent.[95] This is true except for rare, unidentified cases of deep-well injection of wastes.

Table 11
USAGE OF SURFACE AND GROUND WATERS IN SMSAs

Surface H$_2$O (%)	Number of cities	Population (thousand)	Number of cities	Population (thousand)	Average
0—19	26	6,389	65	16,059	247
20—39	6	2,132	15	5,358	357
40—59	6	2,928	15	7,359	491
60—79	7	3,355	18	8,432	468
80—100	66	45,082	166	113,314	683
NUMBER	SAMPLE 111 CITIES		EXTRAPOLATED 279 CITIES		

A. Population Exposed

Table 11 compiles the number of water supplies and populations served by different percentages of surface water (the remaining is ground water) for a sample of 111 U.S. cities. An extrapolation is made to the 279 total existing SMSAs. The largest category of SMSAs primarily uses surface water (80 to 100%) and has an average population of 683,000. The remaining SMSAs have an unweighted average population of 390,000. This is shown on Table 11 as well as in Figure 10 which presents the overall exposure model. The attentuated concentration from the river model, therefore, reaches cities of 683,000 and 390,000 for relatively high surface and ground use, respectively. We assume approximately 80% uptake from the latter to obtain 546,000 surface water users and 20% uptake from the former to obtain 78,000 ground water users as shown in the model.

Cities with high ground water supplies represent 40% of the total cities with the adjusted average population (113 of 279 SMSAs); 60% of these cities are high surface water users. Thus, we adjusted the average size city to contain 360,000 surface water users to account for both ground and surface water usage across all cities.

Based upon this calculation, the maximum total population exposed via surface water contamination is about 100 million (about 2/3 of the population of all SMSAs). This will be important as a population limit in the overlap model.

On this basis, we approximated the concentration and number of exposed people from each plant. When considering total number of plants and concentrations, there is no available information on exposure overlaps. When estimating risks for a linear dose-effect model, such as for cancer, this is not a problem since risk is represented by the number of people exposed at each risk level aggregated additively. However, for chronic exposures, where no-effect thresholds may occur, the overlap of exposure by the same people results in increased concentration and risk nonlinearly.

B. Population Overlap

The total exposure from an industry is simply obtained by multiplying population exposed from one plant by the number of plants. Where the resulting product exceeds 100.4 million, it is divided by the population limit to obtain an adjustment factor for the river model dilution rate. This takes into account the fact that more than 1 plant per industry can be located in a given SMSA (assuming more than 279 plants exist). The population exposed is still 100.4 million, but the exposure level is higher due to resulting lower dilution rates.

C. Industry Overlap

More than one type of polluting industry can also be located in a single SMSA. Thus, overlap of exposure from multiple industries is possible. This is not a problem in determining cancer risks which are assumed to obey a linear dose-effect relation. In this case, each incremental exposure is independently additive to total risk. Thus, exposed populations are directly converted to cancer risks as shown in the upper right hand entries on Figure 10.

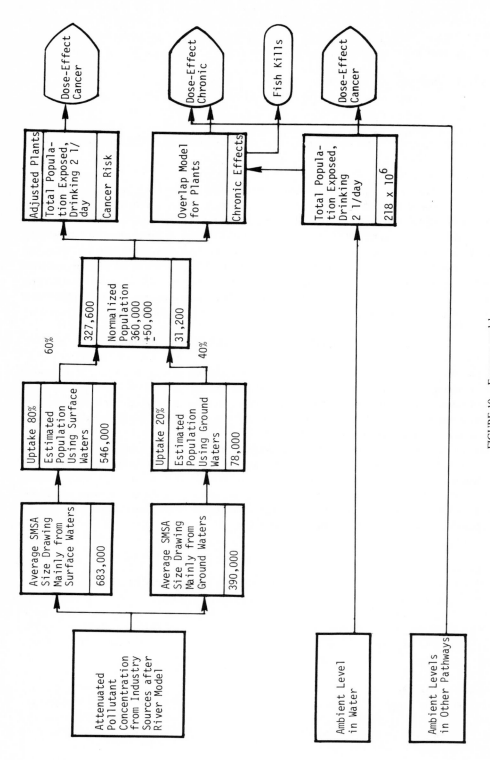

FIGURE 10.   Exposure model

Overlap is important when considering chronic and acute risks since the cumulative effect of multiple exposures is of primary importance. However, that effect is not factored into the equation since the overlap will be small compared to that from ambient levels in water and other media. Thus, chronic levels are estimated as to their particular contribution to chronic exposure only on an industry by industry basis.

D. Ambient Overlap

One must also consider ambient levels in water and in other pathways, as shown in Figure 10. Risks from ambient levels in water are first computed independently to assess their contribution. Then, we estimate the total exposure from all pathways to both general and special populations which provides perspective on cumulative chronic exposure and on the part played by water contamination.

IV. APPLICATION TO CADMIUM

The results of the river and exposure models for cadmium are shown in Table 12. These results will be applied in Chapter 19 where we address cadmium risk to aquatic species and to humans. Again, note that some estimates are derived from the plant data and therefore may not be accurate representations. The electroplating industry affects maximum exposed population because of plant overlap corrections in the river model. The typical measurement threshold is about 1 ppb. Unlike the phenol results, numerous industries have downstream concentrations below this level.

V. APPLICATION TO PHENOL

The results of the river and exposure models for phenol appear in Table 13. The plant direct discharge concentrations (column 3) and the diluted concentrations (columns 5 and 9) are used in Chapter 20 where adverse effects to aquatic species are described. The 2-day attentuation concentrations (columns 7 and 11) and the exposed population also are used in that chapter in determining risk to humans.

The number of plants per industry, the average waste flow per plant, and the average waste concentrations were, in some cases, derived from related information and results may not accurately represent industry estimates. Three industries have more plants than there are total "exposure sets" (279 SMSAs). These industries have the maximum exposed population (100.4 million). Note that 0.10 ppb is the measurement threshold and that most industries presented have higher downstream concentrations.

Table 12
RIVER AND EXPOSURE MODEL RESULTS FOR CADMIUM-EMITTING INDUSTRIES

Industry	Number	Plant Data		Small river/normal (130 MGD)			Average river/normal (17,800 MGD)			Exposed population (millions)
		Avg. waste flow (MGD)	Average concentration (ppm)	Dilution Ratio	Concentration (ppb)	Attenuation (ppb)	Dilution Ratio	Concentration (ppb)	Attenuation (ppb)	
Mining/smelting	7	10.98	15.0	12.8	1171.88	1031.25	1,622	9.25	8.14	2.5
Electroplating										
Job shops[b]	3000	0.038	3.1	319	9.71	8.54	43,564	.07[a]	.06[a]	100
PB manufacture[b]	400	0.019	3.1	4,773	0.65	0.57[a]	653,448	.005[a]	.004[a]	100
Captive shops[b]	6050	0.277	15.0	22.6	662.0	583	2,964	5.1	4.5	100
Pigments	6	0.024	1.3	5,418	0.24[a]	0.21[a]	741,668	0.002[a]	0.002[a]	2
Plastics	193	0.820	0.02	160	0.13[a]	0.11[a]	21,708	0.001[a]	0.0007[a]	70
Ni-Cd batteries	10	0.273	0.77	477	1.61	1.42	65,202	0.012[a]	0.010[a]	3.6
Iron and steel	136	0.785	0.07	167	0.42[a]	0.37[a]	22,676	0.003[a]	0.0026[a]	49
Phosphate fertilizers	92	2.900	16.0	46	347.83	306.09	6,139	2.61	2.29	33

Note: Sources — Census of Manufacturers,[96,99] Yost,[66] Yost and Miles,[69] Patterson.[63,98]

[a] Below detectable limit of 1 ppb (Farnsworth, 1980).
[b] Normalized for overlapping.

Table 13
RIVER AND EXPOSURE MODEL RESULTS FOR PHENOL-EMITTING INDUSTRIES

| Industry units | Plant Data | | | Small River/Normal Flow | | | | Average River/Normal Flow | | | | Exposed population (millions) |
	Number	Average waste flow (MGD)	Average concentration (ppm)	Dilution Ratio	Conc. (ppm)	Attenuation 1 Day (ppm)	2 Days (ppm)	Dilution Ratio	Conc. (ppb)	Attenuation 1 Day (ppb)	2 Days (ppb)	
Coke ovens	38	0.5	3,360	261	12.9	12.3	11.6	35,600	94	89.3	84.8	14
No dephenolization												
With dephenolization	25	0.5	71	261	0.27	0.26	0.24	35,600	2	1.9	1.8	9
Phenol manufacturing	14	1.92	960	69	14.0	13.3	12.6	9,271	104	99	94	5
Petrochemical[a]	453	12.7	367	7.3	50.2	47.7	45.3	864	425	404	383	100
Oil refineries[a]	349	9.8	57	11.6	4.9	4.7	4.4	1,453	39	37	35	100
Plywoods[a]	477	0.053	2.3	1428	0.002	0.002	0.001	195,621	0.012	0.001	0.001	100
Blast furnaces	251	42	0.34	4.1	0.083	0.079	0.075	443	0.8	0.76	0.72	90
Phenolic resins	41	1.9	1,600	69.4	23.1	22.0	20.8	9,368	171	163	154	15

[a] Normalized for overlapping.

Note: Sources — Census of Manufacturers,[96] Patterson[63]

Appendix III-A

PHENOL EXPOSURE ANALYSIS

We have conducted a sample exposure analysis for phenol to demonstrate the types of data and assumptions one can use in estimating exposure. While one may rightfully disagree with some of our assumptions, they are intended only to show the types of thinking that must be done in assessing exposure from all sources.

When actually conducting a risk assessment, studies cited should be scientifically validated for accuracy, good laboratory practice, conclusions drawn, and so forth.

I. GENERAL POPULATION

We will calculate exposure to this population exclusive of presumed higher risk subpopulations. There were approximately 217 million U.S. citizens in 1977. Of these, 158 million live in metropolitan SMSAs.[97] Phenol exposure occurs from several pathways, including water; all of which must be estimated to obtain perspective.

The 202 million U.S. citizens over the age of 5 are here considered to be the general population, since children under 5 may be more susceptible and have different degrees of exposure via the various routes.

We estimated both worst and most plausible case exposure for each source of phenol for each identified group, arbitrarily halving worst case estimates for both dosage received and population exposed to arrive at a more probable situation. Granted, this is a simplistic approach. There are, of course, numerous rationales and ways of arriving at these estimates.

A. Nonwater Pathways
1. Medical Preparations
Table 1-A lists some phenol medical preparations.[86] A worst case situation might consider ingestion or dermal exposure by use of lozenges, mouthwash, medicated cream, liquid cold sore preparations, and rash ointments by some 10% of the population, or about 20 million people over the age of 5 years. These are arbitrary figures. There has been no attempt to quantify actual use.

a. Ingestion
Assumptions

1. Mouthwash — use of 1.45% phenol mouthwash once a day, every day. Ingestion of an estimated 0.5 g of mouthwash per day = 0.007 g/day or 267 g of phenol per year
2. Lozenges — daily ingestion of 8 lozenges, each containing 32.5 mg phenol, used for 1 week at a time, 5 times per year = 0.26 g/day or 9.1 g/year

b. Dermal Contact
We set dermal absorption at 10% to estimate exposure via medicinal preparations. This rate, of course, depends upon the formulation of the product, whether oil, liquid, cream, etc.; percentage of phenol; total composition; duration of exposure; and other factors.

At least one study has measured increased blood levels following dermal application of a product containing phenol.[99] Human volunteers receiving 1 or more applications of 50 g calamine lotion amended with 1 g phenol (2% phenol) applied over 75% of their bodies demonstrated increased phenol blood levels. The level of free phenol increased from 0.15 to about 0.4 mg/100 mℓ while conjugated phenol levels increased from 0.35 to 1.1, 1.65,

Table 1-A
SOME MEDICINAL PREPARATIONS CONTAINING PHENOL

Producer	Medicine	Mode of exposure	Phenol content
Baker Laboratories	Calamine lotion	Dermal contact	1.0%
	Panscol® ointment		1.0%
	P & S® ointment or liquid		1.0%
Glenbrook	Campho-Phenique®, liquid		4.75%
	Campho-Phenique®, powder		2.0%
Fuller	Benadex® ointment		1.0%
Peterson's Ointment Company	Peterson's® ointment		2.5%
Walker Corporation	Dri-Toxen® cream		1.0%
O'Neal, Jones & Feldman	Tanurol® ointment		0.75%
Noxell	Noxzema® medicated		0.50%
Merrell-National	Cepastat® mouthwash	Ingestion	1.45%
	Cepastat® lozenges		1.45%
Eaton Laboratories	Chloraseptic® mouthwash		1.4%
	Chloraseptic® lozenges		32.5 mg per Lozenge

and 1.9 mg/100 ml for the groups receiving 1, 2, 3, or 4 calamine applications, respectively. Both free and conjugated concentrations returned to pre-exposure levels within 24 hr after termination of the application.

Free phenol increased by almost three times and conjugated phenol from three to more than five times depending upon the number of applications. The total increase was 2 mg/ 100 ml of blood.

We used a derived estimate of 7 ml of liquid or 7 g of a cream preparation required to cover a person's hands.[100] The surface of the hands constitute 4.4% of the body surface, or 0.082 m². [101] The adult body surface, 1.85 m², requires 158 ml or grams of preparation to cover, approximately 1 g for each 0.012 m² of skin surface.

Blood constitutes about 7% of human body weight, approximately 5 ℓ by volume for a typical adult male. The increase in phenol, then, for the groups exposed to the most skin applications was about 1.8 mg/100 ml or 90 mg total. Body surface covered (75%) was 1.38 m² and 0.2 g phenol were applied. Four applications of 0.2 g equals 0.8 g or 800 mg. In this instance, phenol absorption was 12%. [99]

The single application resulted in an even higher absorption rate. The blood retained an additional 1 mg/100 ml phenol (50 mg total). One application contained 0.2 g or 200 mg. Phenol absorption was therefore 25%.

Given the unknowns with regard to each exposure, we used a 10% dermal absorption rate, probably conservative, as a worst case estimate.[99]

Medicated lotion

Assumptions
  Use of lotion containing 0.5% phenol
  Used three times a week
  Applied to 1/10 of body surface
  Used over a 6-month period

Calculations
   1/10 Body area = 0.2 m² requiring 17 g of 0.5%
   cream or .09 g
   0.09 × 10% Absorption rate = 0.01 g
   0.01 × 81 Applications = 0.8 g

Liquid preparation

Assumptions
   Use of liquid containing 4.65% phenol
   Used every day
   Applied to 1/20 body area
   Used during 2 1-week periods
   Used over a year

Calculations
   1/20 Body area = 0.09 m² requiring 8 g (or mℓ) of 4.75%
   Phenol or 0.4 g
   0.4 g × 10% Absorption rate = 0.04 g
   0.04 g × 14 Applications = 0.6 g

Rash ointment

Assumptions
Use of ointment containing 2.5% phenol
Used every day
Applied to 1/5 body area
Used during 2 1-week periods

Calculations
   1/5 Body area = 0.4 m² requiring 34 g (or mℓ) of 2.5%
   phenol (0.9 g)
   0.9 g × 10% Absorption rate = 0.09 g
   0.09 g × 14 Applications = 1.3 g

Totals

If one assumes the same 10% of the population over 5 years of age used all the above preparations in amounts indicated, totals for individuals on days in which all preparations are used and yearly totals are

	Daily	**Yearly**
	0.14g dermal	2.7 g dermal
	0.26g ingestion	11.8 g ingestion

If one also assumes use of these products by different, perhaps overlapping, segments of the population, the population at risk would increase and the amount of exposure decrease.

2. Inhalation
   Except for inhalation in enclosed industrial settings, the relatively rapid decomposition in the presence of oxygen makes phenol of negligible adverse consequence to the general population.

3. Consumption of Seafood

Seafood is listed separately for several reasons. These are

1.   Varying residues have been found in a number of aquatic species
2.   Seafood represents a water contamination hazard
3.   Market basket surveys do not include recreationally caught fish, which can represent a sizeable portion of the diet of some groups

There are limiting factors of taste, odor, and tainting which preclude ingestion of seafood contaminated with high levels of phenol. Tainting of the flesh occurs at lower levels than either odor or taste. As a worst case, we assume tainting will not prevent consumption. Taste is affected, depending on species of fish, between 1 and 25 mg/ℓ.

As previously discussed, phenols in both free and conjugated forms are normally found in mammalian tissues; but we found only a few studies concerned with tissues of fish or other members of the aquatic biota, and obtained no information on phenol accumulation in either shellfish or aquatic plants.

Useful studies for estimating exposure from ingesting contaminated finfish are field studies which measure concentration in both fish and water. In such a study, less than 0.3 mg/kg was detected in *Alburnus punctatus* and in roach, *Rutilus rutilus,* from the unpolluted River Isar in Germany; however, there was a concentration of up to 3.2 mg/kg in bream *(Abramis brama)* and barbel *(Barbus barbus)* from polluted parts of the Elbe and Rhine Rivers that contained between 0.2 to 0.7 mg/ℓ phenol.[102] We have not attempted to scientifically validate any studies cited.

Pollution in the Elbe and Rhine was considerably greater than the 11 μg STORET mean of U.S. rivers. In fact, the low German range of 0.2 mg/ℓ is greater than even the highest U.S. river concentration for which we have data, the St. Lawrence (150 μg/ℓ).[93] We selected a level of 3 mg/kg of fish flesh, near the 3.2 level reported in bream and barbel in the German study, as the worst case estimate for 10% of U.S. rivers. Although fish tainting occurs in waters with between 1 and 10 mg/ℓ phenol (the polluted German rivers were in this range), to present a worst case situation we consider this effect acceptable by consumers. In some species, more sensitive for the effect, such as carp at 1 mg/ℓ, taste may also be affected at water pollution levels presumably lower than those resulting in 3 mg/kg in vivo.

The average U.S. household consumes 3.9 lb of poultry and seafood each week.[103] As stated, we have no data on phenol bioaccumulation in either poultry or shellfish.

Assumptions
    3.9 lb Poultry and seafood consumed weekly per household
    25% of 3.9 lb is finfish
    3 Persons over 5 years per household
    3 mg/kg Phenol in finfish

Calculations
    1/4 of 3.9 lb = About 1 lb finfish per household per week
    1 lb/3 = 0.33 lb Per person per week
    0.33/7 Days = about 0.05 lb per person per day
    3 mg/kg Phenol = 0.068 phenol mg per 0.05 lb fish
Each person receives about 0.07 mg/day from this source.

Note that this amount may be greatly increased for those people consuming a great deal more seafood than the average. If, as a worst case, 0.07 mg/day is rounded off to 0.1 mg, it would raise the amount of fish consumed daily to 0.075 lb, still a relatively small amount.

Remember, however, that we are using a worst case residue level. For this reason, we will consider 0.1 mg/day or 37 mg/year as our worst exposure case.

It is, of course, possible to estimate higher consumption levels to raise our worst case exposure if one believes greater exposure could occur.

4. Ingestion (Other than Seafood)

Phenol is a normal metabolic product in animals. We have averaged the residues in 2 reported samples, 7 mg/kg in sausage and 28.6 mg/kg in smoked pork belly, to derive an average of 12 mg/kg for all meat ingested in the U.S.[90] We have no data on phenol in poultry, eggs, milk, or animal fats.

Assumptions (worst case)
   100% Population over 5 years
   10.5 lb/Week consumed by average U.S. household (personal communication, EPA)
   3.5 Persons per household, 3 over 5 years

Calculations
   10.5/3 = 3.5 lb of meat consumed by each person per week
   3.5/7 = 0.5 lb/day
   12 mg/kg or 2.7 mg per 0.5 lb
   2.7 mg phenol are obtained daily

B. Water Pathways

1. Drinking Water

Drinking water appears to be the major phenol ingestion exposure from not only water contamination but all sources, excluding medicinal preparations.

Assumptions (worst case)
   8 Million people of all ages exposed to 1.3 to 1.4 mg/ℓ
   2 ℓ Daily consumption

Calculations
   2 ℓ × 1.3 mg = 2.6 mg/day
   2 ℓ × 1.4 mg = 2.8 mg/day

At the highest level of exposure, 8 million people receive between 2.6 and 2.8 mg each day from this source.

2. Food

The worst case daily dietary exposure excluding drinking water is 0.1 mg for finfish and 2.7 mg for meat and meat products, a total of approximately 2.8 mg/day phenol.

That proportion of dietary phenol resulting from water contamination, the consumption of finfish, is small in relation to "natural" contamination of meat products. Even if the worst case drinking water estimate is added to the amount obtained from eating finfish, the resulting figure is dwarfed by that obtained by consuming meat.

a. Worst Case

There were about 202 million people above the age of 5 in the U.S. in 1977. We estimate 4%, or 8 million people, are exposed to drinking water with a phenol concentration above 1 mg/ℓ. Since this is close to that level where odor and taste become a consideration, it may be presumed that many communities would not long endure a higher level. Dermal absorption during bathing and swimming is presumed to be negligible.

The 8 million people exposed to levels of 1.3 to 1.4 $\mu g/\ell$ receive an estimated 2.6 to 2.8 $\mu g/day$ (in 2 ℓ of water) from this route, or 0.001 g/year. This population would also receive a 2.8 mg/day or 1 g/year from meat and seafood.

If 10% of this population, or 800,000 people, are also exposed to the range of medicinal products available, they could be exposed to relatively high amounts of phenol. Our worst case assumption is 0.4 g/day for days in which the consumer was exposed to all five medicinal products. Although this is somewhat improbable, it is well within the realm of possibility that a person could be daily exposed to an amount in excess of the 0.26 g of phenol in eight lozenges. For the estimated 800,000 people, the small additional loading from other products would probably not produce significant additional exposure.

The estimated 7,200,000 people receiving the high-phenol drinking water level, but not using phenol medicinal preparations, are exposed to concentrations far below those causing adverse health effects. Exposure as a result of water contamination, i.e., the consumption of finfish and drinking water, is relatively insignificant and adds little to total exposure.

b. More Plausible Case

Even by halving the 0.4 g possible daily intake from medicinal preparations to 0.2 g/day, the 2.8 mg/day from diet and drinking water would still be greatly exceeded. There are obviously more sophisticated ways of arriving at a more plausible estimate, but this should suffice for our purposes.

It is obvious that even when we use our worst case drinking water assumptions, the potential for exposure from phenol in medicinal preparations is far greater.

II. POTENTIALLY SUSCEPTIBLE POPULATIONS

A. Children

Fifteen million U.S. children under 5 years of age may be divided into two groups, with different exposure routes, amounts, and probably different responses. Those groups are neonates to toddlers (up to 2-1/2 years of age), and children between 2-1/2 and 5 years of age. We arbitrarily estimate that the population is equally divided. Of this number, only 4%, or 600,000, are estimated to be exposed to public water supplies with phenolic levels between the EPA 1 $\mu g/\ell$ limit and 1.4 $\mu g/\ell$.

1. One month to 2-1/2 years

Population at risk: 300,000 — Drinking water and bathing are probably the only exposure sources for this population. Infant foods, except meats which are usually a minor part of a child's diet, are unlikely to contain phenol, for reasons previously discussed. Dermal exposure from bathing is considered inconsequential. We have no data on the phenol content of mother's milk, the other major food source.

Exposure estimate — Ingest 1.4 mg/day water (assuming 1.4 mg/ℓ worst case water level, and 1 ℓ water consumed daily).

2. 2-1/2 to 5 years

Population at risk — 300,000 for drinking water and bathing; 60,000 for lotion.

In addition to drinking water consumption, arbitrarily estimated at 1 ℓ/day, another possible exposure source is over-the-counter medicinal preparations. At this age, we believe only those preparations for external skin treatment of rash, such as poison ivy, are likely to be used. We assume these products would not be applied to a child's open cut. Dermal exposure from bathing is considered inconsequential.

Exposure estimate — Ingestion of water: as in Group 1, 1.4 mg consumed daily. Dermal contact: the 1.85 m² skin area used in estimating dermal exposure of the general population is for a person 180 cm in height and 70 kg in weight, figures appropriate to the adult male. This, of course, represented an extreme worst case for that population consisting of everyone over 5 years of age.

For children between 2-1/2 and 5, we estimate 1/4 the adult male skin area or approximately 0.5 m².

Lotion contact:

Assumptions
    20% of population exposed (60,000 children)
    Use of product containing 2.5% phenol
    Use daily, 4 times a year, 1 week each time
    Application to 1/2 the body surface (about 0.9 m²)

Calculations
    1/2 Body area = 0.25 m² requiring 21.4 g of 2.5% lotion or 0.3 g
    0.3 g × 10% absorption rate = 0.03 g
    0.03 g × 28 applications = 0.8 g yearly.

3. Conclusions

As with adults in the general population, some children may be exposed to relatively high phenol levels from medicinal preparations.

There are 15 million U.S. children under 5 years of age. Of these, 600,000 are daily exposed to 1.4 μg phenol in 1 ℓ of drinking water. An unquantified, but presumably low, exposure from consumption of meat and meat products must be added. There are 60,000 children between 2-1/2 and 5 years estimated to be exposed to both the highest drinking water levels and lotions containing phenol.

Both the 0.003 g/day worst case and 0.0015 g/day more plausible estimate far exceed the amount likely from drinking water in the worst case scenario.

B. Women

1. In the General Population

Pregnant women may be especially sensitive because of possible fetotoxic and teratogenic effects. Although evidence of such effects are *extremely* weak for phenol, we have discussed the possibility to demonstrate a means of bracketing a worst possible effect. A lack of adequate animal or epidemiological data require that we discuss this effect in the most general terms.

We will assume as a worst case that of the 4.5 million females (those over 5 years of age) exposed to 1 to 1.4 mg/ℓ of phenol in drinking water, 50% are of childbearing age and 20% of these are pregnant in any year. Obviously a ridiculously high estimate. Forty-five thousand pregnant women will be potentially exposed at this level. To this exposure must be added ingestion and medicinal exposure.

If there are minimal or no effects with such implausible input, there should be little question of safety.

2. Occupationally Exposed

Bureau of Labor census and other data on the number of workers in each phenol-polluting industry do not tell us how many workers are actually exposed to phenol. We therefore used a NIOSH estimate of up to 10,000 workers occupationally exposed even though it does not permit differentiation by industry.[78]

Given that some industries producing phenol as a product or by-product are heavy industries, we believe it unlikely that more than 5%, or 500 of the exposed production workers, are women. Even assuming that 1 in 10 of these women workers become pregnant in a given year, only 50 pregnant women will be occupationally exposed annually. Only two of these women will also be exposed to the high drinking water concentration of 1.4 μg/day.

C. Men — Occupationally Exposed

NIOSH has a permissible occupational exposure limit of 5 ppm (19 mg/m³) based upon both inhalation and percutaneous absorption. The latter is the primary source of occupational exposure.[78]

We will assume 100% of occupational exposure is dermal, and as a worse case assume 50% of the 10,000 potentially exposed workers are exposed at least once during the course of a day to the maximum permissible concentration. Although absorption is a function of time, skin area formulation and other factors, we assume 10% absorption as a worse case. On this basis, about 5000 people are exposed to 5 ppm phenol (an absorption of 0.5 ppm) at least once a day. If they come into dermal contact with 2 ℓ of 5 ppm fluid in the course of an 8-hr day, they may absorb 1 mg of phenol.

Although these figures are likely to vary widely because of the nature of phenol, they represent the highest drinking water levels in municipal areas at one point in time. Based on those projections, the worst case estimate is 400 male workers exposed to drinking water levels between 1.3 to 1.4 μg/ℓ and to a total of 1 mg/day on the job.

The occupationally exposed probably, for the most part, live in areas of heavy industry and high population. A worst case situation combining occupation, drinking water, diet, and medicinal preparations could be constructed. However, it is adequate to observe that drinking water and finfish add little additional exposure under any scenario.

Appendix III-B

SPECIAL POPULATIONS

I. CADMIUM

Certain subsets of the general population, because of either an increased sensitivity to the effects of cadmium or an increased exposure, are assumed to have greater risk. Appendix III-A discusses increased exposure. Populations which are potentially more sensitive to cadmium are

1.   Women (especially pregnant women and postmenopausal women with a history of multiple childbirths)
     A.   In the general population
     B.   Occupationally exposed
     C.   Smokers
     D.   Nutritionally deficient
2.   Children
     A.   In the general population
     B.   Nutritionally deficient
3.   Men
     A.   Occupationally exposed
     B.   Smokers
     C.   Nutritionally deficient
4.   Elderly
5.   Industrial/urban populations near sites of contamination
6.   Families of workers
7.   Individuals whose water supplies are soft or acidic

There is evidence linking increased cadmium sensitivity to people nutritionally deficient in zinc, manganese, copper, iron, calcium, proteins, and other dietary essentials. Smokers are exposed to additional cadmium in tobacco.[104]

Women appear to absorb more dietary cadmium than do men.[104] Pregnant women may be even more at risk even though cadmium does not cross the placenta since it accumulates in the embryo of laboratory rats between implantation and formation of the placenta.[110] Another experiment with gravid rats showed 10 mg of orally administered cadmium per kilogram of body weight each day produced fewer implantations and live fetuses as well as more resorbed fetuses.[111] Fetal weight of rats whose mothers were given cadmium in drinking water weighed less than control animals.[112] It also is shown to interfere with fetal uptake of zinc and copper in goats resulting in fetotoxicity. Symptoms of the cadmium induced (described in Chapter 12) Itai-Itai disease in Japan were most evident in postmenopausal women with a history of multiple childbirths who also had calcium and vitamin D deficiencies.[59]

Children have been included as a sensitive population because of evidence, including that cited above, that cadmium retention and adverse effects are greater in younger laboratory animals. It has been estimated that by the age of three a child has accumulated one third his lifetime burden of cadmium.[104] Although the body burden increases with age, the young of all species appear in general to be more susceptible to toxic substances.

Because of the cadmium persistence, families of workers may be exposed to residues on the worker's clothing or person. Populations close to industrial polluting sites are potentially exposed to a greater extent directly from air emissions, the relatively rare spilling or dumping

---

THE MARKET BASKET SURVEY

Infants and toddlers have a greater relative cadmium body intake than do adults.
In 1975, FDA expanded its Market Basket Survey to include analyses of contaminants in infant and toddler foods. The results of the 1979 survey showed that although daily cadmium dietary intake in μg was 5.16, 10.72, and 34.0 for infants, toddlers, and adults, respectively, when corrected to reflect body weight, i.e., the relative intake in micrograms of cadmium per kilogram of body weight, the corresponding values were 0.63, 0.78, and 0.49.[105-108]

Not only is intake greater, but studies with laboratory animals show greater sensitivity and gastrointestinal absorption by young animals.[109]

---

event, consumption of recreationally caught fish and shellfish from polluted water, and swimming in these polluted waters.

A consistent finding in most cadmium-polluted areas in Japan has been the association of increased age with prevalence of proteinuria. This is an indication of renal dysfunction brought on by a cadmium overloading.[104]

Studies have also shown that use of soft water increases the incidence of cadmium-induced cardiovascular disease.[104]

II. PHENOL

Health effects of phenol are similar to cadmium in that a number of subsets of the general population may be at greater risk.

In the case of exposure to phenol, special populations are limited to:

- Those using medical preparations containing phenol
- Children
- The occupationally exposed

Phenol, unlike cadmium, appears to pose no greater risk to either those living in close proximity to industrial sources or to families of workers for the following reasons:

1. Population close to industrial sources
   A. The odor associated with phenolic contaminated water is considered adequate to successfully deter swimming and heavy recreational use of contaminated waters. Waters close to heavy industrial use also generally contain other products from industrial and municipal sources which make these activities unappealing.
   B. Phenol is short lived in the presence of oxygen so inhalation exposure in these areas is presumed no greater than elsewhere.
   C. We have extrapolated contamination of water supplies to the general population. There are too many unknowns to estimate contamination in industrial vs. rural areas.
   D. Exposure to phenol may be greater in rural than urban industrial areas, since those living in rural areas may ingest untreated or partially treated water, while urban residents use treated water. When biological treatment processes are used in urban areas, they are very effective in breaking down phenolic and other organic compounds. The rural population may, therefore, be exposed to potentially larger concentrations if unprocessed rural water supplies (wells, streams, etc.) are contaminated.

2.   Families of Workers
    A. Because of the relatively rapid phenol decomposition, they should be exposed to little or no phenol on the clothing or person of that worker.
    B. Consumption of locally caught seafood would be limited by smell, taste, and tainting of the flesh of fish exposed to low levels.

Human sensitivity to phenol is probably age related. The increased sensitivity of the young of many species to toxic substances, including cadmium as described, has been noted many times. In the case of phenol, 10-day-old laboratory rats were observed to be more sensitive than older rats.[113] Experiments with aquatic species have demonstrated an increased adverse response as the age of the test organism decreased.[114] Although effects in fish or daphnia should not be generalized to man and data from rats are limited, it is reasonable to assume the human population under the age of 5 years may be more susceptible than the general population.

Part IV

DETERMINING DOSE/EFFECT RELATIONSHIPS

The cause is hidden, but the result is well known.

Ovid (Publius Ovidius Naso) 43 B.C.—c. 18 A.D.
Metamorphosis IV, 287

The manifested result of toxic poisoning is termed the *potency* of that toxin. Human and animal studies on our selected pollutants highlight the range and type of subchronic functional disorders as well as total and specific physiological health impacts which may be caused by toxins. Since there are numerous uncertainties in empirical studies with respect to the carcinogenicity, mutagenicity, or teratogenicity of these chemicals, an accurate assessment of these uncertainties is critically important. We must show whether various risk aversion methods have a differential sensitivity within the limits of knowledge.

The selection of our "strawmen" chemicals was based on their contrasting chemical and biological properties in the hope that an analysis of these two pollutants would yield information applicable to a wide variety of water contaminants. There is a voluminous literature on environmental effects and human toxicology of cadmium. The literature on phenol, in contrast, is sparse. This distinction, as well as those listed in Table 14, can be used to determine differences in mechanisms whereby governmental agencies establish standards for pollutants.

In this part, we also estimate the health impact based upon published data.* Our objective is to provide exposure-effect estimates for use in the analysis of alternate regulatory approaches, but no definitive estimates of either cadmium or phenol effects. In each case, we will estimate a range of uncertainty which is retained for the risk estimates. Separate estimates will be made for cancer, noncarcinogenic chronic and acute effects. New information, of course, might require modification of these preliminary dose-response estimates.

Loomis employs a generalized categorization for potency noting that "since a great range of concentrations or doses of various chemicals may be involved in the production of harm, categories of toxicity have been devised on the basis of the amounts of the chemicals necessary to produce harm."[115] This categorization is presented in Table 15 and will be used later to rank doses of cadmium and phenol in an estimated dose-response curve.

Results obtained here will later be used in relating exposure estimates from Part III to health impacts in man and aquatic organisms. All of the dose-effect estimates are uncertain, but we have retained reasonable ranges to show the extent of uncertainty, including no effect conditions. We will examine the sensitivity of standard setting methods for these ranges.

Again, these estimates are based upon sparse information, but as any regulator or "regulatee" must make judgments, so must we. However, one must take care to leave biases out of the estimates. This includes the tendency to be conservative to protect human health which will be considered only when we can see the price which must be paid for this particular bias.

---

* Readers unfamiliar with introductory concepts in toxicology, e.g., acute and chronic toxicology testing and dose response relationships, should consult Loomis, *Essentials of Toxicology,*[115] or Casarett and Doull's *Toxicology*[116] to provide perspective.

Table 14
CONTRASTING CHEMICAL AND BIOLOGICAL CHARACTERISTICS OF CADMIUM AND PHENOL

Contrasting characteristic	Cadmium	Phenol
Chemical type	Inorganic	Organic
Environmental fate	Nonbiodegradable	Biodegradable
Bioaccumulation	Great	Slight
Carcinogenicity	Carcinogen (animal and occupational studies)	Tumor promoter (distantly related compound — in animal studies only)
Mutagenicity	Mutagen (mainly in vitro studies)	Weak mutagen
Reproductive	Teratogen	Nonteratogen
Synergistic/antagonistic interactions	Many environmental variables potentiate or attenuate cadmium toxicity to the aquatic biota; many nutritinal parameters influence cadmium toxicity to animals and humans	Few environmental variables potentiate or attenuate phenol toxicity to the aquatic biota and no nutritional factors are known to influence phenol toxicity to animals or humans.

Table 15
GENERAL CATEGORIZATION OF TOXICITY

Extremely toxic	1 mg/kg or less
Highly toxic	1 to 50 mg/kg
Moderately toxic	50 to 500 mg/kg
Slightly toxic	0.5 to 5 gm/kg
Practically nontoxic	5 to 15 gm/kg
Relatively harmless	more than 15 gm/kg

Chapter 12

CADMIUM POTENCY (FOR MODEL CALCULATIONS ONLY)

*Perfect health, like perfect beauty, is
a rare thing and so, it seems, is perfect disease.*

Peter Mere Latham (1789-1875)
Collected Works, Blk. 1, Ch. 173

Ascribing a cause and effect relationship to a toxin affecting a given person moving among many toxins and other health-affecting causes is difficult. The lack of a perfectly distinct manifestation of a particular disease attributable to a single cause does not help. In measuring potency, therefore, we often use animal laboratory tests and extrapolate the results to humans. Study of cadmium poisoning is no exception.

I. ACUTE EFFECTS

Environmental health assessments have been made of direct cadmium toxicity from dietary or inhaled sources,[59,60] while others have developed system-level analysis methods to quantify human exposure levels.[69]

Cadmium was first isolated as a sulfide in zinc ore in 1817. While zinc is a trace nutrient essential to cell metabolism, cadmium has no known metabolic function and is extremely toxic to living matter. Animals have a limited defense against small amounts via a metal binding protein. This inactivates free cadmium for a time and accounts for the slow manifestation of functional disorders from lower cadmium levels.[66]

Cadmium is a cumulative poison. Whether it enters the body through contaminated water, food (grown where cadmium-containing phosphate fertilizers or sludges were applied), or air (from coal furnaces, zinc smelters, or refuse incineration), excessive build-up over a period of 20 to 30 years can eventually cause serious liver and kidney damage.*[59,60,69]

Here we will consider effects from both inhalation and ingestion of cadmium. Table 16 compares chronic and acute cadmium induced functional disorders. We will begin our discussion with a consideration of acute effects. Additional test results of subcutaneous injection and oral administration of various cadmium salts are presented in Appendix IV-A.

A. Inhalation

Inhalation may result in pulmonary distress or failure without any definite or immediate symptoms. An acute inhaled dose is considered highly toxic and may, after 4 to 8 hr, result in pulmonary edema with symptoms ranging from a metallic taste in the mouth to the usual "flu" symptoms.[118]

Inhalation risk is greatest when cadmium is in a fresh fume. Arc-generated fumes are considered more toxic than those thermally generated.[119] Cadmium salts, inhaled as dusts or aerosols, may also be quite toxic. Lethal doses are expressed as the air concentration in milligram per cubic meter multiplied by exposure time in minutes. For thermally generated fumes, the lethal dose is considered to be about 2800 min-mg/m^3 (280 mg/m^3 for 10 min); for arc-generated fumes, about 1400 min-mg/m^3 (140 mg/m^3 for 10 min).[120]

---

*   For a more exhaustive coverage of respiratory, renal, histopathlogical, and biochemical effects caused by acute cadmium toxicity, the reader is directed to *Toxicology of Metals*, Vol. I, March 1976, U.S. Department of Commerce.[117]

Table 16
COMPARISON OF CHRONIC AND ACUTE CADMIUM-INDUCED FUNCTIONAL DISORDERS

Acute exposure-ingestion	Chronic exposure-ingestion
Vomiting	Mild anemia
Gastrointestinal distress	Enteropathy and cellular changes in intestinal mucosa
Headache, vertigo	Damaged renal tubules
Exhaustion, shock — death	Osteoporosis (poorly mineralized bones)
Renal failure with anuria	Prostate cancer (in some cadmium exposed workers)
Renal failure with anuria and uremia — death	Hypertension (in some cases)
Acute exposure-intramuscular injection	**Chronic exposure-intramuscular**
Inflammation	N/A
Necrosis	
Acute exposure-inhalation	**Chronic exposure-inhalation**
Pulmonary edema	Perivascular and peribronchial fibrosis and emphysema
Death by anoxia	Disturbances in absorption of zinc, iron, copper, and calcium
Alveolar cell proliferation	
Hyperplasia of the lining cells	
Intra-alveolar hemorrhage	

B. Ingestion

Food poisoning has occurred when organic acids in food or beverages react with cadmium-plated containers to form organic cadmium salts, in turn producing poisonous cadmium chloride when combined with gastric juices. Because of numerous incidents of such food poisoning prior to 1945, the FDA has adopted a policy of refusing use of cadmium-containing materials in either food packaging or preparation.[121] Table 17 (after Fulkerson and Goeller)[60] provides an approximated scale of acute toxicity. Ingesting 3 to 15 mg is reported to have produced vomiting. Given equal concentrations in food and water, water is considered the more toxic since some components of food produce antagonistic effects and vary in their chemical binding with cadmium.[94]

II. NONCARCINOGENIC CHRONIC EFFECTS

Cadmium is one of 18 metals classified as a moderate to severe industrial hazard, causing a number of adverse chronic health effects (see Table 16).[122] The effect manifested depends upon the compound, dose duration, and route of exposure.

Again, we will discuss health effects from the two major routes of entry of this metal, inhalation and ingestion. Appendix IV-A presents additional toxicologic evidence not directly used in conducting our assessment.

A. Inhalation

Proteinuria and emphysema may occur after a latent period following chronic inhalation of cadmium dust or fumes.[60] The precise amount of exposure which induces proteinuria is unknown. The earliest manifestation of chronic cadmium exposure has been studied more in laboratory animals than in humans where observations are limited to industrial exposure.[59] Although levels of 200 ppm wet weight in kidney cortex are associated with proteinuria,

Table 17
TOXICITY SCALE FOR INGESTED
CADMIUM

3 to 90 mg	Emetic threshold reported nonfatal accidents
15 mg	Experimentally induced vomiting
10 to 326 mg	Reported severe toxic symptoms (but not fatal)
350 to 3500 mg	Estimated lethal dose
8900 mg	Reported lethal dose

Axelsson and Piscator did not observe the disorder at levels of 440 ppm.[123] Based on the former, if human gastrointestinal absorption is approximately 3%, proteinuria requires an ingestion rate of 500 to 600 µg.

Smoking may account for the above differences. According to Friberg et al., smoking 20 cigarettes per day results in the inhalation of about 2 to 4 µg of cadmium.[59] Cadmium inhaled in cigarette smoke is absorbed at a much higher rate than that ingested because food may slow absorption. Lewis et al. autopsied corpses of men who had smoked one or more packages of cigarettes per day and found cadmium concentration in the kidney to be twice that of comparable nonsmoker control groups.[124] Similar studies confirm that smoking 30 or more cigarettes per day doubles the body burden derived from food intake alone[125] (See Appendix IV-A). About 16% of the cadmium resides in the ash, 15% in filters, and 69% in smoke.

B. Ingestion

Oral routes of exposure from food and water produced a severe disease in Japan. This disorder, termed "Itai-itai byo" (ouch-ouch) disease was first noted about 10 years after the probable initial contamination in 1924 resulting from zinc-lead mining in the Jintsu River basin. It was most common among postmenopausal women who had borne many children (an average of six).

Local inhabitants had ten times the usual daily intake of cadmium by Japanese and manifested symptoms which included pains in the back and legs, pressure on the bones, skeletal deformities, a ducklike gait, increased susceptibility to multiple fractures after trauma as slight as coughing, renal pathologies such as increased protein (proteinuria) and glucose (glucosuria) in the urine, abnormalities of the gastrointestinal tract, and anemia. Cadmium was not officially recognized as the cause until 1968, after more than 100 deaths had occurred. Diagnosis was complicated by the presence of zinc and lead which also have toxic effects and the marginal nutritional condition of the victims.[59]

Other studies have shown symptoms of chronic poisoning to generally be relatively mild. They include anemia, enteropathy, damaged renal tubules, and osteoporosis.[104] The onset of these disorders may be linked to, or accelerated by, disturbances of iron, copper, zinc, calcium, or manganese metabolism. Deficient intakes of these trace elements greatly increases cadmium toxicity in laboratory animals.[70] Table 17 presents a toxicity scale for ingested cadmium (after Fulkerson and Goeller).[60]

III. CARCINOGENIC POTENTIAL

The following summary of delayed toxicity is based in part on the previously cited EPA Cadmium Position Document which includes an assessment of oncogenicity by the EPA Carcinogen Assessment Group (CAG).[126]

A. Carcinogenic Effects

After a review of oral and injection route animal studies and human inhalation and dermal contact epidemiologic studies, CAG concluded "that (i) the number of tumors at sites removed from the site of injection increased significantly in animals injected with cadmium; (ii) significant increases in prostate cancer were found in persons occupationally exposed to cadmium and (iii) chronic oral studies, although reported as negative by their authors, must be considered inconclusive because of protocol deficiencies."

The CAG also noted that some mutagenic actions of cadmium, including mammalian cellular transformations, have a high correlation with oncogenicity.

The EPA position document cited studies reporting a statistically high incidence of prostate cancer in exposed workers. In areas with greater than 3 $\mu g/\ell$ of cadmium in water, there was a positive correlation with increased cancer of the larynx, esophagus, intestine, pharynx, lung, and bladder; but because these sources presented no evidence either for or against cadmium complicity in these other cancer types, the definitive degree of cadmium carcinogenicity must be considered uncertain for our purpose.

B. Mutagenic Effects (As They Relate to Carcinogenicity)

As previously noted, CAG states that cadmium is a potent mutagenic agent. However, correlations between mutagenicity and carcinogenicity remain unconfirmed. Cross-species comparisons are "fundamentally incommensurable test systems".[127] This criticism applies to many studies in addition to those for cadmium where carcinogenic tests are conducted on rodents while mutagenic studies are conducted on bacteria or fruit flies.

IV. SUMMARY — CADMIUM EFFECTS

Although cadmium is known to be highly toxic to animals and humans, the extent of placental transfer, the exact rate of body accumulation, and the precise levels associated with hypertension and prostate cancer are still disputed in the published literature.

The EPA Cadmium Working Group determined that cadmium exceeded EPA risk criteria for mutagenicity, teratogenicity, and fetotoxicity. In addition, they cited a number of possible adverse effects to the lungs, kidneys, liver, carbohydrate metabolism mechanism, gonads, and the cardiovascular, skeletal, hematopoietic, and central nervous systems.

Functional disorders and health effects from different exposures are summarized in Figure 11. We assume levels of exposure below approximately 10 ppb to be background levels. These concentrations are widely found in both U.S. surface waters and normal human blood. Exposure levels from 10 to 100 ppb correspond to residues in the blood and urine of exposed workers as well as in food and milk samples from a wide range of geographical and institutional settings. Exposure over a 30-year period to 100 ppb to 1 ppm, corresponding to soil, sediment, and crustal abundance concentrations, can induce kidney damage. Values above 1 ppm have been linked to chronic diseases of the blood, liver, and kidney in rats and humans. Levels above 10 ppm have been associated with rapid onset of acute illness in children, and chromosomal and skeletal abnormalities in laboratory rats and mice.

These summary values are taken from various published sources which suggest the absence of a clear threshold for the various adverse effects assuming body retention.[60,69,104,115,128] The independent responses to different concentrations and dosages for various cadmium salts, routes of exposure, and test animals is plotted on a log scale in Figure 12, producing a linear no-threshold pattern. It is evident that metabolic excretion occurs, but at very low levels. Thus, a threshold may exist if excretion is not proportional to tissue concentration.

Estimated dose-response relationships for noncarcinogenic chronic poisoning are shown in Figure 13. Ranges of uncertainty for dose-response points indicate orders of magnitude of uncertainty.

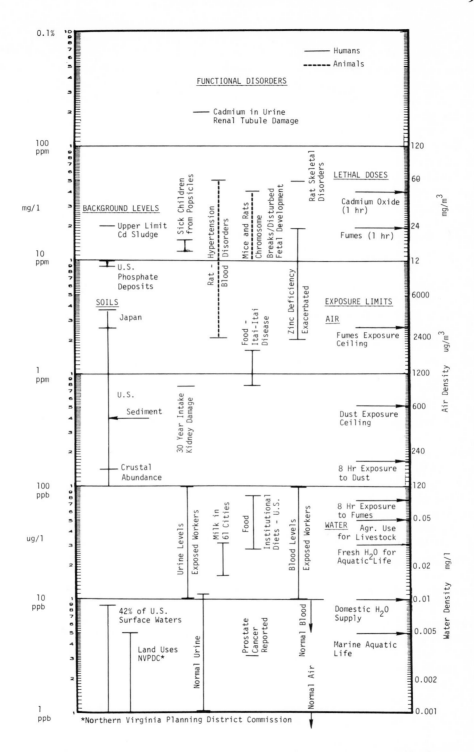

FIGURE 11. Overview of cadmium-induced functional disorders by organ system with relevant benchmarks.

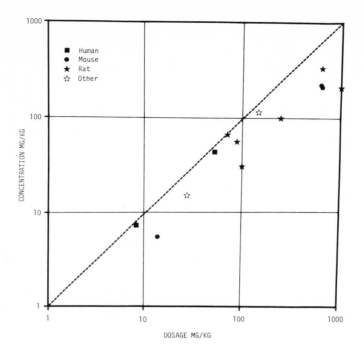

FIGURE 12.   Concentration vs. dose for cadmium salts on selected species.

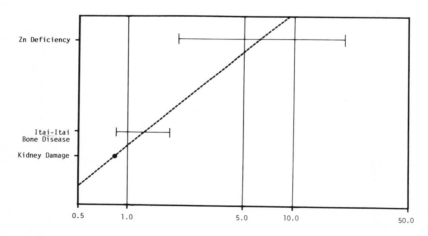

FIGURE 13.   Noncarcinogenic cadmium chronic dose responses for humans (estimated).

V. RANGE OF UNCERTAINTIES

Data obtained have various degrees of uncertainty associated with estimation. To the extent meaningful, we presented these uncertainties as obtained.

For example, lifetime human exposure is uncertain. Using statistical modeling techniques, the Stanford Research Institute (SRI) has determined excess deaths per 100,000 exposed people based on a range of exposure values (see Figure 14).[64] We assume these figures represent a reasonable range of uncertainties. Other effects or factors listed here likewise have associated uncertainties.

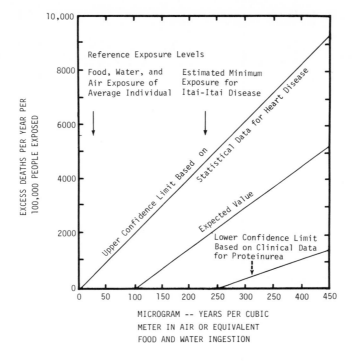

FIGURE 14. Cadmium cumulative lifetime exposures vs. increased mortalities.

A. Hypertension

Hypertension in humans has been statistically linked to long-term cadmium intoxication. However, because these studies are believed inconclusive,[59] additional epidemiological evidence is needed to demonstrate that exposure to cadmium compounds causes cardiovascular disease in humans.

B. Cancer

Animal malignancies can be produced via injection of cadmium salts. In addition, some data suggest a higher incidence of prostate cancer in exposed workers.

C. Excretion Rates

Cadmium urinary excretion rates are very low while the range of gastrointestinal excretion needs further investigation.

D. Metabolic Interactions

The poor nutritional regimes of the Japanese Itai-itai victims clearly contributed to the severity of the poisonings. Evidence linking cadmium effects with zinc and calcium metabolism as well as both liver and kidney disorders appears to warrant further research attention.

Much of our knowledge of cadmium dose-effect relationships is based on observations of occupationally exposed workers. Different segments of the general population may, of course, have varying sensitivity to cadmium.

Because substantial amounts of cadmium appear in urine only after renal tubule damage and because blood levels reflect only recent exposures, better methods are needed to detect chronic human poisoning.[104] Relatively minor changes in the dietary levels of zinc, manganese, and copper dramatically influence the effect of cadmium on human health.

Chapter 13

CADMIUM DOSE-EFFECT RELATIONSHIPS
(FOR MODEL CALCULATIONS ONLY)

Truth in all its kinds is most difficult to win;
truth in medicine is the most difficult of all.

Peter Mere Latham (1789-1875)
Collected Works, Blk. I, .60

We believe there is only inconclusive evidence that cadmium is a carcinogen. There are, however, other renal and cardiovascular impacts of poisoning by this metal which appear to cause more immediate health problems. At worst, cadmium may be a carcinogen of low potency. For example, comparative evidence suggests it to be less potent than arsenic. Therefore, we have used estimates of arsenic carcinogenicity to provide one rational assumption for an upper limit for cadmium dose-effect relationships. This chapter also includes estimates of chronic human effects and effects on aquatic organisms.

I. ACUTE EFFECTS AND NONCARCINOGENIC CHRONIC EFFECTS

Noncarcinogenic cadmium poisoning in humans results from ingestion via food and water pathways. As described, annual intake of 0.9 to 1.7 ppm for a 30-year exposure period can result in kidney damage and the Itai-itai disease syndrome where an estimated 50% of the exposed population became ill. This range is shown on Figure 15 as entry A. The abscissa shows two scales for noncarcinogenic chronic poisoning. The upper scale represents the annual intake for 70 years and the lower, the total intake over 70 years, assuming total retention. The latter permits comparison with acute exposure as shown at the top of the figure. The ordinate shows an approximate scale of percent disease occurrence in humans. At 200 ppm, renal tubal damage occurs and it is assumed that 75 to 100% of the exposed population became ill. This is shown as entry B on Figure 15. Exposed workers had blood and urine levels of 0.01 to 0.1 ppm with no observable health effect as shown in entry C. A reported general population value of 0.016 ppm in the kidney results in no observable effect (point D). Also, reported values of 25 ppm in smokers (entry E) and 50 ppm in those having other diseases, generally fell within a 75 to 90% range of occurrence. The latter two entries are encompassed within the range of uncertainty shown graphically through upper and lower estimates. Entry D is outside the range indicating that the no effect level may be underestimated. We used reported ranges of cadmium poisoning from published literature in these estimates.[59,60,104]

Cadmium poisoning may produce noncarcinogenic chronic effects in both vertebrate and invertebrate marine, estaurine, and freshwater species. As with phenol, the severity of effects appears to depend on a number of environmental factors as well as the age of affected organisms. There are many subtle reproductive, growth, and survival effects which are difficult or impossible to translate into overall impact or survival.

Lethal effects on finfish generally occur between zero and 10,000 $\mu g/\ell$ depending upon species, exposure period, and environmental factors.[129]

The range of chronic effects used here are data selected from the published literature as representative of "worst case" chronic effects. For vertebrates, the range used as the span of these effects is 3 ppb (reduced growth of brook trout, *Salvelinus fontinalis*)[130] to 10 ppm (reduction of egg production by zebra fish, *Brachydario rerio*).[131] Note that even though the subtle effects of reduced growth on survivability of members of the affected species and

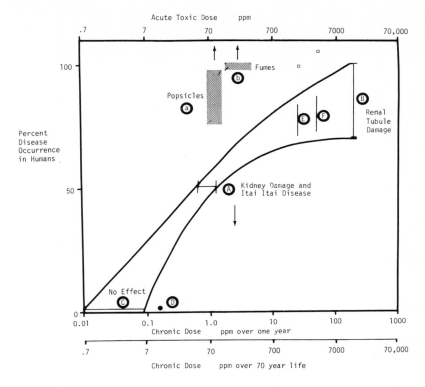

FIGURE 15. Acute and noncarcinogenic chronic cadmium human health impacts. A, well-documented episode; B, AURA estimated range; C, well-documented no effect level.

OTHER CONSIDERATIONS FOR POTENCY DETERMINATION

There are a few things to keep in mind when considering dose/response extrapolations. One is that dietary factors alone, such as the food itself, fungi, or the cooking process, can be a risk. We are all aware of the concern that high dietary fat intake may cause a build-up of cholesterol leading to heart attack. Amazingly, there is also some evidence that protein deficiency may, under certain circumstances, actually decrease tumor production by blocking the body's B-lymphocyte cells from producing antibodies, thus *protecting* the cancer cell. Many other factors, such as viruses, radiation, and cigarettes may be synergistic in producing adverse effects.[21]

The tumors may result from the action of initiator, promotor, or co-carcinogenic substances. The first must be present for cancer to develop and appears to cause cellular mutations even with a one-time exposure. The second often must be present for an effect to be manifested (this exposure is long term) and the third functions as an initiator or promotor only when another co-carcinogen is present. The effect of promotors, but not of initiators can be reversed. With few exceptions, we do not know the classification of carcinogenic substances.[21]

on the ecosystem are difficult, if not impossible, to establish, this range is chosen as a worst case to illustrate the method we will follow. This is a relatively simplistic approach. We have, for example, made no attempt to determine either the most sensitive species or the value and sensitivity of the species we have chosen.

For invertebrates, a range of 1 ppb, which reduced number and size of broods of *Daphina pulex,*[132] to 10 ppb which reduced viable oyster larvae,[130] will be used.

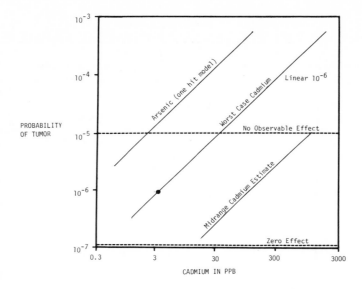

FIGURE 16. Alternate estimates of the probability of human prostate cancer over lifetime (70 year) exposure.

School children were acutely poisoned when they ingested popsicles contaminated with 13 to 16 ppm cadmium. We assume that 75 to 100% of those exposed became ill. This is shown in Figure 15 as a shaded area showing the ranges of uncertainty (entry A), but using the abscissa scale at the top of the figure. Lethal doses were 20 and 45 ppm for cadmium and cadmium oxide, respectively, from inhalation of fumes. These levels resulted in close to 100% lethality (shown as a shaded rectangle in entry B). In addition, prolonged exposure to cadmium dust produced emphysema in exposed workers.[59]

The bottom scale of Figure 15 approximates the lifetime intake at the annual chronic levels shown in the scale just above it. Although we did not factor in excretion, the figure permits comparison of that lower chronic scale with the acute level scale of single dose intake at the top. Evidently, cadmium is more toxic for an equivalent intake at the acute exposure level than for chronic exposure. Single large doses are, therefore, more toxic than equivalent doses administered over a long period.

II. CANCER

For the purposes of this study, we considered three alternatives:

1. A potency equivalent to a low estimate of arsenic potency
2. Zero effect
3. A level of 3 ppb, which is assumed to equal 1 in a million (1×10^6) risk likelihood for human testicular cancer for a 70-year intake at this level, as a single value for downward extrapolation on a linear basis

In the latter case, the data come from the single report in the literature for prostate cancer in human males.[126,133] The 3 ppb level at which the cancer was reported is extrapolated downward linearly such that a tenfold reduction in concentration results in a tenfold reduction in lifetime risk. This is illustrated in Figure 16 as a worst case assumption. It is based upon our single point extrapolation on a log-log scale.

The arsenic data is estimated from risk levels for arsenic in an EPA Criteria Document

Table 18
SUMMARY OF HUMAN HEALTH IMPACT OF CADMIUM

Direct toxicity		Long-term health effects	
Acute	**Chronic**	**General**	**Carcinogenic**
Minimum	**Minimum**		
Ingested 3—90 mg	Ingested .087—1.3 mg/day (50 yr—70 kg man)	Possibility of being weak mutagen but studies inconclusive	Sarcumata induced when rats injected with cadmium compounds
Inhaled 0.25 mg/m³ (8 hr)	Inhaled 0.013 mg/m³/day (50 yrs—70 kg man)	0.01—0.02 mmol/kg of Cd Salts in male mice caused neurosis of testes; however study inconclusive	Tumors found mostly in tissues of mesenchymal mesodermal
Severe	**Severe**		
Ingested 10—326 mg	Ingested 0.6—5.25 mg/day (50 yr—70 kg man)	2 mg/kg Injections of Cd sulfate in golden hamster were teratogenic; however, study inconclusive	No cancer induced when mice ingested
Inhaled 1 mg/m³ (8 hr)	Inhaled 0.1—5 mg/m³/day (15—40 yrs—70 kg man)	Reduction of effects can be achieved by adding vitamin C, zinc, sulfhydryl compounds, and selenium	No conclusive evidence that cadmium exposure results in higher incidence of cancer than norm
Lethal	**Lethal**		
Ingested 350—3500 mg	Ingested 0—5.25 mg/day (50 yrs—70 kg man)		
Inhaled 2.9—6.1 mg/m³ (8 hr)	Inhaled 0.17—6.6 mg/m³/day (50 yrs—70 kg man)		

establishing ambient water quality criteria.[134] This extrapolation, based upon three points from that reference, uses the same slope as that document. These data are based upon a modified ''one-hit'' model where 0.0002 corresponds to a 1 in a million lifetime cancer risk. Cadmium appears to be over two orders of magnitude less toxic than arsenic. (The arsenic data are also shown in Figure 16.)

Zero effect is a distinct possibility (Figure 16) since tumors were not observed above levels equivalent to 900 ppb. This may be explained by either one of two hypotheses, namely, that cadmium is not carcinogenic or that cancer is masked by other chronic diseases at these higher levels. Since cancer is age dependent, life-shortening by acute or other chronic toxicity might effectively mask the onset of cancer. With this latter hypothesis in mind, we made an arbitrary mid-level estimate two orders of magnitude lower than the worst case assumption. A tabular summary of human impact is given in Table 18.[59,60,104]

Chapter 14

PHENOL POTENCY
(FOR MODEL CALCULATIONS ONLY)

Every cause produces more than one effect.

Herbert Spencer (1820-1903)
Essays on Education (1861)
On Progress: Its Law and Cause

Phenol, C_6H_6O, also known as carbolic acid, is a weak acid, highly soluble in water and common organic solvents. As noted previously, it is formed in bacterial decomposition of plant and animal wastes and is also found in certain living aromatic plants.

Compounds containing phenol produce a variety of health effects, depending upon condition of the test species and exposure route. For example, phenol appears to be more toxic to certain aquatic organisms, such as mollusks, where it interferes with respiratory gas exchange on the gills, than it is to humans. Figure 17 compares toxicities to various test species with levels found in humans and notes the maximum levels for drinking and irrigation water.

I. ACUTE EFFECTS

Acute response varies both among and within species. This response variability is seen among humans. There are reported cases of human survival well above exposures to 500 mg/kg and of death at exposure levels well below 500 mg/kg. Among animals, rabbits, for example, tolerate exposure levels of 50 to 100 mg/kg,[135] which is in the lethal range for cats.[136]

The key issues in interpreting acute phenol reactions are the compound used, dosage, route of administration, and the age, health, and general vigor of the test animal as well as the sensitivity of the detection method, if we attempt to relate body levels to effects. We have used levels reported from numerous sources in preparing Figures 17 and 18 to present a rough pictorial view of phenol toxicology.

II. NONCARCINOGENIC/CARCINOGENIC CHRONIC EFFECTS

EPA set regulatory exposure limits based upon detectable odor in domestic water supplies at 1 ppb (Figure 17) in 1975.

Fish show pathologic changes in gill tissue when exposed to phenol levels between 20 and 75 ppb and fish and oyster egg development is affected when levels reach 1 to 10 ppm. (Within this range, free and conjugated phenol is produced in human sweat). Background levels from 10 to 100 ppm have been reported for human urine and feces.

Persons exposed via contaminated well water to levels of about 100 ppm for 1 month following a spill at Lake Beulah, Wisc. showed temporary illness including diarrhea, sores, burning of the mouth, and darkened urine.[74] For perspective, the above exposure is comparable to that of people ingesting 8 Chloroseptic®* lozenges a day during the same 1-month period! (See Appendix IV-A.)

Phenol is the base compound for a wide variety of industrial and pharmaceutical products. The risks of exposure to these compounds tend to vary as indicated in Appendix IV-A. There appear to be minor human chronic health effects from these compounds.[137]

---

*   Trade name for a nonprescription throat lozenge containing phenol.

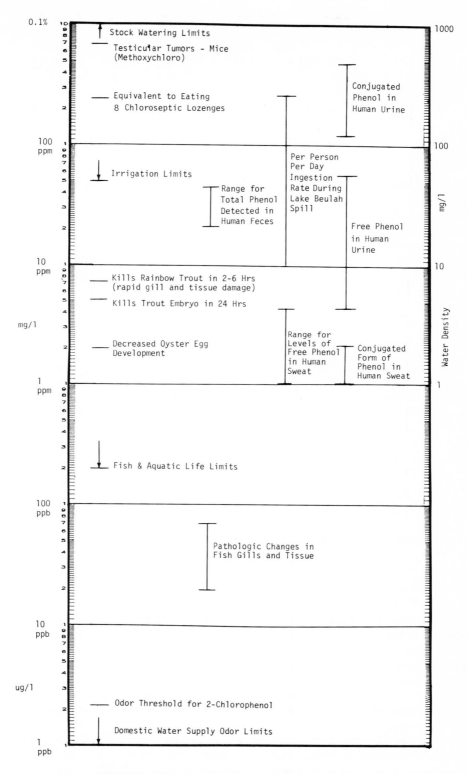

FIGURE 17. Overview of phenol toxicity and health effects.

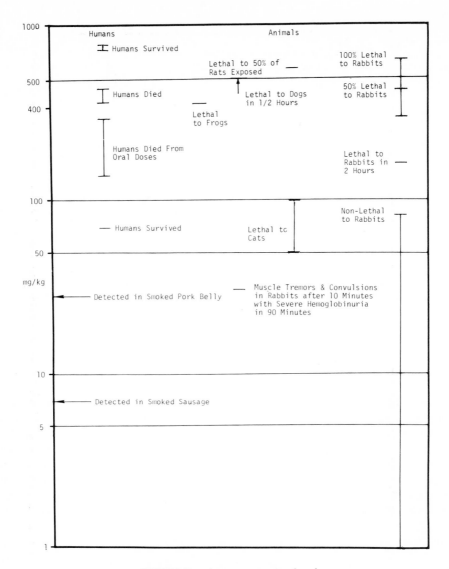

FIGURE 18.   Acute response to phenol.

The potential of pure phenol for mutagenic, oncogenic, teratogenic, and fetotoxic effects is less than that of cadmium. Phenol, even after chronic exposure, does not accumulate in laboratory rats.[138]

Despite this, a number of hazardous phenolic compounds were selected for a 1977 National Academy of Sciences study on drinking water and health based on meeting at least some of the following criteria:[137]

- Experimental evidence of toxicity in man or animals including carcinogenicity, mutagenicity, and teratogenicity
- Identified in drinking water at relatively high concentrations
- Molecular structure closely related to that of another compound of known toxicity
- Component of a pesticide in heavy use that could result in contamination of drinking-water supplies

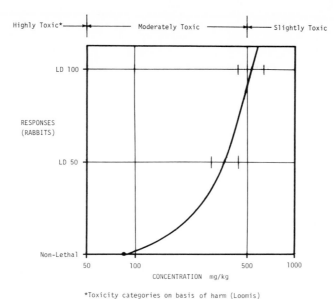

FIGURE 19.   Acute phenol dose response curves (estimated) for rabbits.

- Listed in the Safe Drinking Water Act or National Interim Primary Drinking Water Regulations

We found the quantity and quality of toxicological information on these compounds to be, in general, very uneven. Numbers do not imply a guaranteed level of safety. Rather, they indicate that exposure to a single compound at a given level of exposure has produced the recorded health effect. Although the compounds studied are suspected of producing carcinogenic effects (otherwise, they would not have been chosen for this study), this carcinogenic potential has not, as yet, been established in man. In Figure 17, we have used the carcinogenic effect of a very distantly related compound, the insecticide methoxychlor. This is discussed elsewhere. Note that the Acceptable Dietary Intake (ADI) levels set by FDA for toxic substances in food do not account for other potential nonwater routes of contamination.

III. SUMMARY — PHENOL EFFECTS

During internal cellular movements of toxic substances, metabolic processes may transform phenol into a toxic material.[115] This would help account for variation in dose response for similar individuals or for the same individual at different times. On the other hand, biological systems can metabolize some low-level or occasional toxins until the systems are overwhelmed.

Risk assessments based on high dose exposure extrapolations must take into account certain key variables, several of which are currently unavailable for phenol. The most critical are:

- Differential sensitivity by species or animal strain
- Variation in tissue susceptibility, pharmacology and substance kinetics
- Biotransformation
- Bioaccumulation patterns

Dose-response curve estimates in Figures 19 through 21 are based on information which

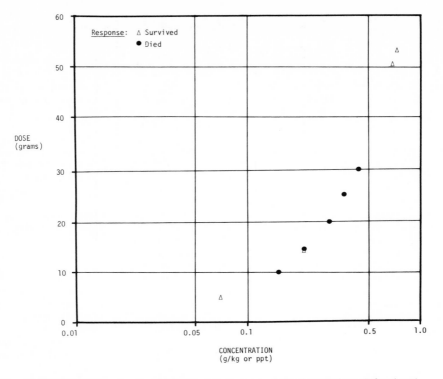

FIGURE 20. Human acute and chronic dose response relationships (estimated) for phenol.

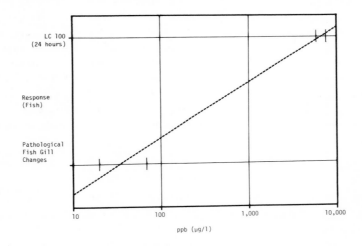

FIGURE 21. Chronic phenol dose response curve (estimated) for fish.

was presented in Figure 18. These values, when compared with the categorization of toxicity in Table 15 (after NAS, 1977[137]), support the conclusion that phenol is only moderately toxic to mammals but extremely so to certain fish.

IV. RANGE OF UNCERTAINTIES

The ranges of uncertainty for phenol are complicated because of this lower human toxicity and because it appears to produce less severe human and animal health effects than its many

Table 19
HEALTH EFFECTS FOR PHENOLIC COMPOUNDS

Compound	Uses	Critical levels	Health effects
Methoxychlor (half-life 46 days)	Insecticide 10 million lb/yr	Rats LD_{50} 6000 mg/kg Mice LD_{50} 2900 mg/kg Monkeys LD_{50} 2500 mg/kg 750 mg/kg — 2 yr mice via diet	Low mammalian toxicity; testicular tumors in some mice strains
2.4 Dichlorophenol (up to 36 $\mu g/\ell$ detected in U.S. drinking water)	In organic synthesis	36 $\mu g/\ell$ highest concentration detected in finished H_2O Mice LD_{50} oral 1.63 g/kg Rats LD_{50} oral 4.5 g/kg	312 mg/kg — 39 weeks Promoted papillomas and carcinomas in mice
2.4 Dimethylphenol (slightly soluble and found in finished water)	Coal tar derivative. Many industrial uses[a]	Mice oral LD_{50} 809 mg/kg Rats oral LD_{50} 3200 mg/kg Mice topical LD_{50} 1040 mg/kg Mice intraperitoneal LD_{50} 150 mg/kg	Topical carcinogenic; uncertain long-term effects
O-Methyoxyphenol (detected in finished water in the lower Mississippi)	Medicines	Rabbit LD_{50} 3.74 g/kg Rats — lethal 50 mg/kg Rats LD_{50} oral 725 mg/kg Mice lethal 0.4 g/kg Mice 0.15% solution Mice 0.6% solution No data available on chronic effects, mutagenicity and teratogenicity Rats augments the carcinogenicity of tobacco smoke	Unknown for humans. Rats — disorders of the testes and germinal epithelium Mice — paralysis of heart Mice — paralysis of smooth muscle Cattle — hemolytic action on blood plus irregular RNA synthesis, protein synthesis, and some interference with mitochondrial respiration Sheep — inhibition of fatty acid; oxygenase Rodents — leukopenia and leukocystosis
Pentachlorophenol (PCP) (up to 1.4 $\mu g/\ell$ found in U.S. drinking water)	Wood preservatives	Human — minimum lethal 29 mg/kg Mouse — LD_{50} oral 120—140 mg/kg Rat — 27—100 mg/kg Guinea pig — 100 mg/kg Rabbit — 100—130 mg/kg Dog — 150—200 mg/kg	Human — loss of appetite, respiratory difficulty, anesthesia, hyperpyrexia, sweating, dyspnea, and rapid progressive coma (Disagreement over interpretation of subacute and chronic toxicity tests)

[a] Phenolic antioxidants, pharmaceuticals, plastics, resins, solvents, disinfectants, insecticides, fungicides, rubber, chemicals, wetting agents, dye stuffs.

(Source: NAS[137])

metabolities. Phenol may, in fact, pose greater but unknown risks to health and the environment through the decomposition or combustion of end use products than through water pathways.

The range of health effects for rats, mice, sheep, and cattle exposed to o-methoxyphenol, a phenol metabolite (Table 19), points to a potential uncertainty in test procedures and the possibility of investigation differences. Did all investigators look for all organ system functional disorders? How much variation exists among members of the same species? Is age a critical unrecorded factor in animal responses to chemical agents? These are but a few of the uncertainties in the use and interpretation of secondary sources.

Chapter 15

PHENOL DOSE-EFFECTS RELATIONSHIPS
(FOR MODEL CALCULATIONS ONLY)

Find out the cause of this effect, or rather say,
the cause of this defect, for this effect defective
comes by cause.

William Shakespeare
Hamlet (1600—1601), Act II

In this chapter we will calculate dose-effect ranges for the carcinogenic potential of phenol. Unlike cadmium, chronic phenol effects to humans appear limited to carcinogenicity. We will also estimate adverse effects to aquatic species.

I. ACUTE EFFECTS

Humans have died from oral phenol doses between 200 to 350 ppm. Acute illness and flu-like symptoms resulted between 1 and 3 months after a tanker spill resulted in contamination levels of 10 to 280 ppm. Nevertheless, phenol in free and conjugated forms is reported in human sweat and urine with no observed health effects at ranges of 1 to 60 ppm.[86] These ranges are displayed in Figure 22 via a linear central extrapolation and a worst case envelope. Human health effects are summarized in Table 20.[137,139] There is evidently a threshold for acute effects somewhere between the 1 and 10 ppm level (if not higher).

Acute toxicity for fish, as derived from the published literature, ranges from an LC_{50} of 9.4 for rainbow trout to 63 ppm for mollies. For invertebrate aquatic organisms, the range is from 14 ppm in crustaceans to 780 ppm for clams (some shellfish are more resistant). Thus, acute toxicity for aquatic species encompasses the range for humans on either side.

II. NONCARCINOGENIC CHRONIC EFFECTS

In humans, only acute exposure produces known effects. Aquatic organisms, on the other hand, are affected by, and generally exposed to, low levels of phenol in aquatic environments. They may, however, be exposed to higher concentrations. Salmonid fishes (rainbow trout), for example, were shown to avoid phenol contamination at either the chronic or acute levels.[140] Since the odorless level is well below chronic effect levels, humans, who would presumably shun odoriferous water, generally are not exposed at the chronic level.

Chronic effects in fish, as described in Appendix IV-B, would probably result between 0.1 and 10 ppm depending on species and water condition factors. At the lower levels spawning might be affected. Death might occur at the higher levels.

III. CANCER

Phenol has not been reported to induce cancer. The compound methoxychlor (2,2-bis-(p-methoxyphenyl)-1,1,1-trichloroethane), however, in one FDA laboratory test, produced a statistically significant increase in testicular tumors in 1 strain of mice at 750 ppm but not in another strain tested at the same time[141] (see Table 19).

Methoxychlor is an insecticide, chemically similar to DDT. Although a far more complex

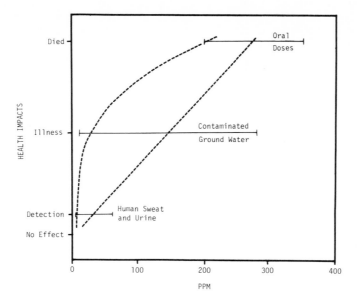

FIGURE 22.   Acute phenol human health impacts (assumed for total population).

molecule than phenol, methoxychlor contains a phenol moeity.* We have no evidence that the presence of this moeity, or the fact that methoxychlor is relatively rapidly detoxified to mono- and bis-phenol derivatives in mammalian systems, account for the carcinogenic potential of methoxychlor. Despite this, we will make a great leap and use this methoxychlor data to predict phenol carcinogenicity. This is done not only to illustrate how one may proceed when faced with a small amount of evidence of carcinogenicity, but also to enable us to use phenol as an appropriate test vehicle for this study.

Using the single positive value, methoxychlor cancer values were scaled to a range for humans using factors of 10 (the EPA recommendation for scaling) and 30 (the NAS recommendation).[137] Linear extrapolation on a log-log scale is shown in Figure 23 for both estimates. Comparing these extrapolations to those for cadmium in Figure 16 (Chapter 13), phenol, if a carcinogen at all, is at least 30 to 100 times less potent. For purposes of this exercise, we will assume phenol and chlorinated phenols, resulting from reaction of raw phenol to chlorination of drinking water, to be similar in potency to methoxychlor (itself either a weak carcinogen or noncarcinogenic). Both assumptions are carried throughout the risk estimates.

---

*   A portion of its molecule.

Table 20

SUMMARY OF HUMAN HEALTH IMPACT OF PHENOL[98,137]

Level	Direct toxicity		Long-term health effects	
	Acute	Chronic	General	Carcinogenic
Minimum	Ingested 0.9—2% concentration Skin 10% solution	Ingested 10—240 mg/ℓ per person per day	Medicinal preparation; discoloration of skin with 10% phenol used as a disinfectant	No human cancers were reported from pure phenol
Severe	Ingested 15—100% concentration	Ingested	Phenol does not accumulate in the body even after chronic exposure	Phenol derivative methoxychloro produced testicular tumors in mice at levels of 700 ppm
Lethal	Ingested 15—100% concentration 29 mg/kg	Ingested 10—240 mg/ℓ per person per day	Diarrhea, mouth sores, darkened urine	

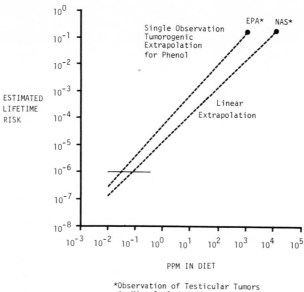

FIGURE 23.   Estimated human tumor probabilities for phenol.

Appendix IV-A

CADMIUM HEALTH EFFECTS/ADDITIONAL INFORMATION

I. ACUTE EFFECTS

A subcutaneous injection of cadmium into the testes of rodents produced cardiovascular effects such as capillary stasis, interstitial edema, severe hemorrhages, altered hemoglobin levels, and necrosis or tissue death. These responses were dose-dependent and reversible. The lowest dose for the most sensitive mice was 1 mg/kg. No testicular tissue death has been reported in occupationally exposed workers.[126]

Acute cadmium poisoning may result from intake of various amounts of different cadmium salts. The broad toxic dose range (LD$_{50}$) of these salts to rats, 72 mg/kg cadmium oxide to 1225 mg/kg cadmium stearate via the oral route, suggests that the particular salt to which an organism is exposed may be toxicologically significant. However, when doses are converted to ingested equivalents, the range narrows considerably and the variations appear less significant.[60] The Cadmium Working Group of the Office of Pesticide Programs, EPA, has decided that "regardless of the cadmium salt used, the results observed can be attributed to the cadmium ion".[126]

The lowest toxic dose causing adverse effects (TDL) of cadmium sulfide, which is mined with zinc, used extensively in pigment manufacture, and relatively insoluble, has been measured at 110 mg/kg. On the other hand, both cadmium chloride, used in preparing CdS pigments and very soluble, and cadmium oxide, used as a stabilizer in certain pigments and relatively insoluble, are toxic at much lower doses; 2.2 mg/kg and 12 mg/kg, respectively.[126]

II. NONCARCINOGENIC CHRONIC EFFECTS

The Carcinogen Assessment Group (CAG) also reported evidence of emphysema and kidney damage in humans following long, low dosage occupational exposure. Studies cited by CAG show cadmium interference with liver functions, including glucose metabolism and insulin levels, of chronically exposed workers. Acute doses by injection caused blood pressure increases and hypertension in rodents. Although the findings are disputed, some studies showed hypertension in humans from areas of the world where total cadmium body burden is high.

CAG found that cadmium did not cause bone marrow changes nor did it accumulate in bone tissue, but it has affected metabolism of both calcium and phosphorous. Thus, effects on those with Itai-itai disease included decalcification and deterioration of the joints and painful pressure on nerves. These health impacts were complicated by low protein diet and vitamin D deficiency.

Finally, chronic exposure produced vertigo, headaches, and changes in neuromuscular reflexes among exposed workers. Other reported central nervous system effects included olfactory disturbances, anosmia (the loss of sense of smell), and hyposmia (reduced sense of smell) in 50% of the workers exposed to 3 to 15 mg/m³ over a 10-year period. Experimental animal studies indicated that rats injected with 0.25 and 1 mg/kg/day for 45 days accumulated brain residues resulting in decreased brain cell activity as measured by concentrations of norepinephrine, acetylcholine, and 5-hydrotryplamine.[126]

CAG cited evidence that cadmium interferes with the replication of genetic material causing point mutations in DNA during replication of chromosomes in bacteria, yeast, and Chinese hamster embryo cells. For example, cadmium chloride produced mutations in bacterial test systems, and cadmium sulfide produced chromosome breaks in cultured human leukocytes, white blood cells. These same chromosomal aberrations were detected in white blood cells of the Japanese victims of Itai-itai disease. Cadmium chloride also changed chromosome number in the ovaries and testes of mice receiving acute doses of 3 or 6 mg/kg. CAG concluded that cadmium meets the required EPA multi-test evidence criteria as "a potent mutagenic agent".

Cadmium is a potent animal teratogenic and fetotoxic agent. Of 24 studies reviewed by a CAG contractor, "none of the studies...established a no observable effect level."

As we have discussed, cadmium does not appear to be transported across the placental barrier but teratogenicity is quickly manifested if the placental exposure occurs at a critical state in embryonic development.

Fetotoxic effects to mice exposed on the 7th day of gestation included reduced fetal weight, exencephalia, abnormal nucleic acid synthesis, and reduced brain growth. Embryos of mice exposed on the 14th to 17th days of gestation died from hemorrhaging. These effects are also seen in other mammals, including rats and hamsters. Women occupationally exposed delivered smaller than average babies and had a mean placental cadmium level of 74.5.

Targets in animals and humans include the gastrointestinal tract, respiratory, central nervous, circulatory, and skeletal systems and, in the excretory system, the kidneys. The kidney cortex usually contains the highest concentration; however, with high exposures, the liver may also accumulate excess metal. A comparison of acute and chronic induced functional disorders is shown in Table 16.

Cadmium is excreted via the urine and feces as well as in lost hair. Estimates of the half-life in the human body range from 18 to 38 years.[116] EPA has estimated that it would require 50 years for the critical level to be reached in the liver:

"It has been estimated that with 5 percent gastrointestinal absorption, rapid excretion of 10 percent of the absorbed dose, and 0.05 percent daily excretion of the total body burden, it would take 50 years with a daily ingestion of 352 μg of Cd to attain the critical level of 200 ppm wet weight in the renal cortex."[94]

Appendix IV-B

PHENOL ACUTE WILDLIFE EFFECTS/ADDITIONAL INFORMATION

Phenol is a neurotoxin. Its effects are accentuated in water by reduced oxygen and by increased salinity, hardness, and temperature. Young fish or those lacking certain nutrients are most affected. Fish kills from phenol are hard to discern since other pollutants are often present, and phenol decomposition occurs relatively rapidly.[114]

Sublethal concentrations cause a series of pathological effects, including necrosis and increased mucus production of the gills,[131] decreased activity of alpha-amylase in the hepatopancreas,[142] destruction of erythrocytes,[143] and histopathological changes in heart, liver, spleen, and skin tissues.[144]

Oocyte damage was noted in rainbow trout *(Salmo gairdneri)* exposed to a concentration below the 48-hr LC_{50} value of 7.5 mg/ℓ.[144] Egg production in the American oyster *(Crassostrea virginica)* was decreased in the presence of 2 mg/ℓ phenol.[145] Guppies kept for a year in 12.5 mg/ℓ phenol initially spawned at the age of 5 months compared to 10 months in the control; however, phenol-exposed guppies exhibited decreased sexual activity.[114]

Apparently some species of fish may be subjected to relatively high concentrations. Bluntnose minnows *(Pimephales notatus)* could detect phenol at concentrations below 0.01 mg/ℓ and distinguish between phenol and *o*-chlorophenol.[114] Yet when given the choice of clean water or water amended with phenol, the minnow, *Phoxinum phoxinus,* did not choose clean water in preference to either 4 or 400 mg/ℓ phenol. The fish entered the phenol-containing solution even though, in such solutions, they lost their sense of balance and capability of coordinated movement.[146] Similarly, rainbow trout did not distinguish clean water from that contaminated with 0.001 to 10 mg/ℓ, even though the latter concentration was lethal.[140]

PART V

ESTABLISHING BASELINES

The least initial deviation from the
truth is multiplied later a thousand-fold

Aristotle (384 to 322 B.C.)
On the Heavens, bk. 1, ch. 5

Data gathering in Part III produced information on, among other things, industrial production and employment and possible alternative compounds or products. Now, using that data, we will establish a *baseline* providing:

- Information for formulating types of standards
- A reference point against which we can evaluate different methods of establishing levels in standards (using the same data base)

In this part, we also will wed dose effect relationships (Chapters 13 and 15) to our exposure model (Chapter 11) in order to establish a *risk baseline* from which risk from new emitting plants can be estimated. Inaccurate or faulty baselines may, of course, result in erroneous standards. Here our baselines were developed only to illustrate the process, not to reflect absolute accuracy.

We have already reviewed the most important industries from the standpoint of water contamination. In Part V, probable primary and inadvertant cadmium and phenol emissions from these industrial sources, as well as the costs of substitute technologies, economic impacts, water quality improvements, and risk from existing exposure are determined. Much of the information on industrial processes, markets, products, and substitutes used in Chapters 16 and 17 and Appendices V-A and V-B was obtained from confidential company sources.

Note that our first concern is adverse effect on human health, the second biota impact, and the third a consideration of these in respect to economic effects.

Since all methods use the same data, it is sufficient that the baseline approximates the actual environmental and economic conditions well enough to allow comparison of methods; it need not exactly depict actual conditions. We reiterate that our risk estimates should not be interpreted as a necessarily factual representation. Emphasis is on methodological development rather than definitive and exhaustive assessment of the pollutants. While our estimates may be imprecise, they nonetheless provide a gross approximation of the major human exposure pathways.

One must consider substitutes not only as to their economic feasibility and quality, but also as to the potential effect of each upon the environment. For instance, a substitute might be more harmful to ecological balance than the original product, as with some early phosphate detergent substitutes. One problem encountered by regulators is that alternatives to the investigated uses of a suspected pollutant cannot, because of time and financial constraints, be explored to the same degree of detail as the latter.

It is relatively easy to measure direct industrial costs in terms of the additional expenditures to reduce pollutants to meet regulated levels. However, evaluation of even these tangible costs is complicated by the effects of inflation on the value of the dollar, the significance of absolute cost levels on industries of different sizes (economies of scale), the degree of competitiveness within a given industry, and the significance of capital costs vs. operating costs.

Indirect factors are more difficult to evaluate. In extreme cases, the costs of pollution control may precipitate economic collapse of marginal firms with resultant loss of jobs — a result some might be unwilling to accept in return for clean water and reduced exposure to toxic substances. Other firms may raise prices, passing cost to the consumer, resulting in lost purchasing power.

The rate of capital obsolescence can increase with high technology processes to meet regulatory stipulations. This has already happened in some industries. Pollution control costs may also remove funds available for other investments, impacting long-term productivity. In addition to all this, there may be a reduced competitiveness in foreign markets of costlier American goods.

The size of an industry is an important variable. Effects on a small industry, such as rubber reclamation, would obviously not have as severe an impact on prices, production, and national income, as would effects on a large industry, such as automobile manufacturing.

We know at this point in our analysis that cadmium and phenol are very different in their uses, risks, and means of control. Although the basic model used is identical in both cases, discussion is tailored to each substance, stressing key considerations of each.

Chapter 16

VALUE OF TECHNOLOGY — CADMIUM

*It is easier to discover a deficiency in individuals,
in states, and in Providence, than to see their
real impact and value.*

Georg Wilhelm Friedrich Hegel
Philosophy of History, 1832

To this, we might add, "and sometimes in industry."

To provide a comparative perspective on the impact of approaches to establishing standards, one must first evaluate the beneficial use of the regulated substance. In addition, we estimate cadmium concentration and costs of reducing this concentration in industrial effluents, as well as identify and assess potential substitutes. We must consider goods inadvertently containing cadmium and services, such as water treatment plants, which deal with it. Commercial value is determined by importance to the general economy and includes product quality and safety as well as employment factors.

In this section, the focus is primarily on the direct industrial producers and users, although the inadvertant industrial consumption and the impact of cadmium contamination on waste-disposal processes are considered. Where appropriate, substitutes are briefly discussed in terms of quality, safety, toxicity, and economic feasibility. They do not, however, represent an exhaustive search. We did not, for example, search for experimental substitutes. Table 21 summarizes the major industrial uses, the value of each industry, the end sectors, and the substitutes. Appendix V-B contains more detailed information on these alternatives.

I. MINING AND SMELTING OF ZINC AND CADMIUM

Cadmium is derived primarily from copper, lead, and (especially) zinc ores. The geographic mining patterns for zinc indicate that the eastern U.S., defined as that area east of the Mississippi River, has predominantly zinc ores, the area west of that river has generally mixed or polymetallic nonferrous ores.[60] As one might expect, the majority of zinc production is in this eastern region.

The cadmium/zinc ratio is much higher in the south central and western than in the eastern states. Total reserves and resources indicate that there is much more total cadmium in the western region. In fact, approximately 70% of domestic cadmium production comes from 4 states west of the Mississippi River.[60] With total 1979 U.S. production at 2270 kkg*, the amount produced from these 4 states is approximately 1589 kkg. There are about 4600 workers in the zinc industry, approximately 3600 of whom work in production.

Cadmium is mined from ores containing other marketable and valuable materials. Zinc is only one of many. Substitution of cadmium in the various uses we will be discussing, therefore, should not impact cadmium effluents from mining and smelting of zinc. The primary industries for production or use of cadmium and cadmium compounds and relevant information on these industries appears in Table 21.

II. ELECTROPLATING

Of the approximately 20,000 electroplating shops scattered throughout the U.S., 10%

*   1 kkg = 1000 kg = 1 metric ton

Table 21

PRIMARY INDUSTRIES IN PRODUCTION/USE OF CADMIUM AND COMPOUNDS

Industry	Number of plants	Employees		Dollar value × 10⁶	Major end uses	Substitutes
		Total	Production			
Cadmium mining and smelting	7	>4,600	~3000	~16	50—55% Electroplating 12—13% Pigments 12—20% Plastic 5—22% Ni-Cd 2—8% Miscellaneous	
Electroplating Anti-corrosive, resists chemicals, less bulky, low electrical resistance	20,000 (10,000)	193,200	(~60,000)	9,410.4 (1,223.4)	48—55% Transportation 26% Electronic communications 19—25% Miscellaneous hardware	Zn, Pb, Al, Cr, Ni, Cu, Sn. Bulky and other than 2n, no electro-chemical protection
Inorganic pigment True yellow to red colors, low solubility, light solubility	105 (6)	11,800	7,900 (<1000)	1,630 (163)	75% Plastics 10% Artist colors 8% Coatings/inks 7% Miscellaneous application	Hydrated ferric oxides Lead chromate Synthetic ferric oxides Lack stability and brilliance
Plastics manuf. Plastics products Used as pigment and heat stabilizer	424 10,043	59,200 460,000	38,200	14,700 30	25.8% Packaging 20.0% Building and construction 10.3% Consumer and institutional 8.4% Electrical and electronic uses 8.4% Adhesive-links countings 8.3% Miscellaneous 7.4% Exports 4.8% Furniture 7.1% Machinery	Pigment as above Heat stabilizer Calcium-zinc Butyl-tin Octyl-tin Relatively more expensive and lower performance rate than barium-cadmium

Ni-Cd battery good performance, highly recyclable	10	3600	2600	104	Calculators 90% Tools, Radios, Telephones, Emergency lighting, 10% Aircraft	Lead-acid for conventional uses Edison cell for heavy duty poor low temperature performance, low life
Alloys	NA	NA	NA	NA	Low melting-point Electrical contact Transportations Fire safety Radiators	
Industries inadvertently containing cadmium and compounds						
Phosphate fertilizers	92	16,000	12,000	3,700	Crops-feed grains, wheat soybeans, cotton tobacco	
Secondary metals industry						
(1) Nonferrous	NA	19,300	~16,000	3,598		
(2) Iron and steel	504	441,900	350,500	41,998	Steel products, cars	

represent over 33% of the capacity of the industry. The majority of these high-production plants are located in the Northeast.[60] The leading states are New York, Michigan, Illinois, and Ohio, followed by California and Texas.[96]

The estimated output value of the electroplating industry is $9410.4 million. About 13%, or $1223.4 million, of this total output value is attributed to cadmium plating.[65,96] There are approximately 193,200 production employees in the metal finishing industry of whom 60,000 may potentially be involved with cadmium plating.

The electroplating industry has three major parts:

1. *Job shops* — about 5000 plants with a total value of $2.1 billion and total production water use of 95 MGD*
2. *Printed circuit board (PB) manufacturers* — about 400 plants. Only PB plants are significant users of metal finishing processes which have a total value of $610.4 million and total production water use of 7.5 MGD for electroplating (of a total 8.7 MGD)
3. *Captive sector* — about 6080 plants with a total value of $6.7 billion and a total production water use of 1700 MGD. The total number of captive shops is about 13,000 plants, but only 47% are considered significant metal finishing operations.[147]

Most cadmium-plated products are used by the transportation industry (48 to 54%). Electronic communications use about 26% of these products and the remainder is used for various hardware requirements.[60,65,66,148] Cadmium electroplate thickness is approximately 0.0007 cm, equivalent to a deposition of 60 g/m^2.[65] Approximately 2140 kkg of cadmium was consumed by the industry in 1979.

Growth in cadmium use by the electroplating industry is estimated at 2.2% annually. The major advantages of cadmium as a protective coating are that it provides:

● Easy, rapid, and uniform deposition even on intricately shaped objects
● High ductility, allowing plated parts to be stamped or formed
● Good electrical conductivity
● Excellent anti-corrosive and soldering properties
● Protection by a relatively thin cadmium coating

Table 21 lists some possible substitutes, but, as detailed in Appendix V-B, where substitutes are feasible in terms of quality and safety, they have for the most part already been employed. Since cadmium is used primarily in products demanding high quality and safety, such as aircraft, some automobile uses, and electronic communication equipment, it is relatively price-inelastic. Estimates of cost increases to over $25/kg (currently about $7/kg) are not expected to seriously affect demand in the electroplating industry. However, EPA believes that increasing costs, whether by tax, tariff, or effluent regulation, could result in 20 to 21% closures with the loss of 12,600 jobs and $335 million in sales.[147] Although a serious ripple effect might be expected throughout the economy, we have not located studies of the total impact.

III. PIGMENTS

Cadmium is used for producing pigments in the yellow to red range for coloring plastics, paints, enamels, inks, ceramics, and glass. Cadmium sulfides are used for colors in the yellow range, cadmium sulfides and selenides for the red orange, and combinations of cadmium compounds for the orange range. Other compounds produce greenish yellows and maroons; even some rich brown colors incorporate cadmium pigments.[60]

* MGD = million gallons per day

Approximately 75% of all cadmium pigments are used by the plastics industry. The remaining uses include artists colors, coatings (primarily used for automobiles), and specialty inks (primarily used for coloring Formica®).

Total worth of the inorganic pigment industry was $1630 million in 1979. The plastics industry consumes approximately 10% of the inorganic pigments. Imports of the latter have increased greatly since 1974, and now approach 3 times the amount of exports. The entire industry employs approximately 11,800 individuals of whom over 1,000 are in the cadmium pigment industry. Approximately 150 of these are engaged in production. Of the 150 inorganic pigment plants, 6 manufacture cadmium pigments.[96]

Increases in the value of the entire inorganic pigment industry appear due more to high prices than to sales. Real growth is not expected to exceed 2%/year for the period up to 1984.[149]

Cadmium pigments have the advantage of low solubility, low cost, excellent hiding power and color intensity, and good heat resistance. Both inorganic and organic pigments can be used as substitutes but present certain problems.[60]

In general, it appears that replacement of cadmium pigments with substitutes depends upon the specific use. Most economic and noneconomic substitutes require color sacrifice. Conversely, those satisfying color requirements are either toxic or expensive. There is additional information on pigments in Appendix V-B.

IV. PLASTICS

Cadmium is used in the plastics industry in heat/light stabilizers and pigments. In fact, 75% of the pigments used in this industry contain cadmium. Barium-cadmium stabilizers, with 7% average cadmium content, are the type of stabilizer most frequently used.[65]

One of the major growth areas in plastics is the pipe and fittings market (plumbing uses) with approximately 15% of the total market.[149] PVC is the preferred material for use in most plastic pipes and constitutes the major use of barium-cadmium stabilizers. Not only do these stabilizers have better performance, but they are also less expensive than calcium-zinc, butyl-tin, and octyl-tin stabilizers. Barium-cadmium stabilizers account for 50% of the stabilizers used in thermo plastics. With these stabilizers, PVC is impervious to heat and ultraviolet light. They may be used to make both rigid and flexible PVCs. Despite a ban on cadmium stabilizers in plastics which come in contact with food, the use of barium-cadmium stabilizers is one of the fastest growing cadmium uses. As yet, however, they comprise only 1 to 2% of the stabilizer industry.[150] Again, the substitutes, lacking in many desirable characteristics, are listed in Appendix V-B.

Although growth may dip in the short term because of the dependence of the industries on petrochemicals, overall it may be expected to return to 8 to 10%. Value of the plastics industry in 1979 was $14.7 billion with a value of imports of $530 million and exports, $2565 million.[149] Total 1979 production was 42 billion lb; PVC production was approximately 6.3 billion lb.

The packaging industry is the largest market for plastics, followed by the building and construction, then consumer and institutional sectors. Transportation uses, still only 5.7% of the total, are increasing significantly — up 60% in 1978 from 1975. There are, of course, many other markets including electronics, adhesives and coatings, furniture, machinery, and export. Although production and consumer costs are going up in the plastics industry, it should not retard growth since the cost of competitive materials is increasing even more rapidly.[149]

In 1977, there were 424 plants in the industry, 337 with 20 or more employees. The major producing regions are the southern, western, and east north central states. Total employment in 1979 was about 59,200 with 38,200 working in production. Because this industry is capital intensive, the labor force is relatively stable with a 1% annual growth.[149]

The plastic products industry, comprising both independent plastic processors and plastics materials and captive processors, is $20 million. There are 10,043 plants of which 4363 have 20 or more employees. Total employment for reporting industries is 460,000 people.[149] These figures, although representative of growth, greatly underestimate the total since recent statistics have not been reported for captive processors, which constitute a substantial portion of industry employment.

V. NI-CD BATTERIES

The nickel-cadmium storage battery industry is one of the fastest growing segments of the cadmium industry with an estimated 17% annual growth rate.[151] Currently, there are only ten U.S. plants manufacturing these batteries.

The 1977 industry value was $103.8 million of which $76.4 million, or 73%, was devoted to sealed sintered plate batteries.[96] Total employment is estimated at 3600 of whom 2600 were production workers.

Nickel-cadmium storage batteries are robust and much longer lived than lead-acid batteries. They also perform better under a range of temperatures, operating at both high and low temperatures and providing maximum current delivery with low voltage drops and low rates of self-discharge. These batteries, however, are more expensive to purchase than conventional batteries. The Ni-Cd industry producing these batteries is an excellent recycler and conserver of cadmium, due in part to the high concentration in wastewater.

The Edison® Cell, composed of iron caustic potash and nickel oxide, is a potential substitute for heavy duty use, but does not perform well at low temperatures and has a higher rate of self-discharge. The lead-acid battery is a suitable substitute for some uses, but, in addition to being much shorter lived, can have toxic effects.

VI. ALLOYS

Cadmium is used in conjunction with bismuth, lead, and tin to achieve a low melting point (45°C), critical in applications such as fire detection apparatus, fire door release links, and safety plugs for compressed gas cylinders and tanks. This is a "minor" use of cadmium, but while dollar value of the industry is not a significant portion of GNP, the nonquantifiable value in terms of risk to human life is important.

Although not an extensive use, it may also be used in conjunction with nickel as a bearing alloy. As a brazing alloy, it is combined with silver, copper, and zinc, allowing joining of ferrous and nonferrous metals. Finally, copper cadmium alloys: now used in place of copper in automobile radiators.

Of the alloy applications, the low melting point alloys have virtually no acceptable substitutes. Substitutes either contain lead, which is more toxic than cadmium; are rare and thus more costly; or exhibit lower performance. Melting points range from 15.7°C to 254.5°C which may suffice for some uses but not for the safety industries, such as fire apparatus manufacturing. On the other hand, there are acceptable substitutes for soldering and brazing. These are based on a silver-copper-zinc alloy with the possible addition of nickel or tin. Cadmium-zinc alloys used for soldering aluminum also appear to have an appropriate substitute in tin-zinc alloys.[60]

VII. PHOSPHATE FERTILIZERS

Most phosphate fertilizer production is in the seacoast regions of the U.S., specifically Florida, Texas, California, and Louisiana. Plants in the southeastern states contribute about 84% of the U.S. production; western sources the remainder. The western ore is considerably higher in cadmium content (40 to 980 ppm vs. 2 to 25 ppm in the Southeast).[152] Phosphate fertilizers derived from western ore contain 27 to 175 ppm cadmium compared to 5 to 14 ppm in fertilizers made from southeastern ore.

Our most recent information valued the phosphate fertilizer industry in 1979 at $3.7 billion with a 1980 increase of 23% to $4.55 billion.[153] Growth is a function of increase in both grain prices and crop acreage in production. Exports in 1979 comprised about 31% of the U.S. phosphate fertilizer industry value.

There are 92 phosphate fertilizer producers in the U.S. Of these, 68 firms employ 20 or more employees. Total employment is 16,000 of whom 11,000 work in production. Over 3×10^6 kkg of phosphate fertilizers is nationally distributed to about 94×10^6 ha yielding feed grains, wheat, soybeans, cotton, and tobacco.

Among industries inadvertently releasing cadmium to the environment, this is the most important to our analysis in that residues are significant and fertilizer use is widespread. Not only is monetary value of the industry high ($3700 million), but, since it is applied to basic U.S. food crops, phosphate fertilizer prices directly impact food prices and potential export crops. Raw products for the manufacture of phosphate fertilizers are abundant.

VIII. SECONDARY METALS

The secondary metals industry includes smelting and refining of both ferrous and non-ferrous metals. The ferrous industry is primarily iron and steel; nonferrous industries include copper, zinc, nickel, magnesium, and tin.

The total industry value in 1977 was $3598 million which has probably not significantly changed in real money terms.[96] Illinois, Pennsylvania, California, and New York together account for almost 40% of the approximately 19,300 workers in this industry. The secondary metals industry is a relatively minor cadmium polluter because of metal recycling. The primary environmental concern here is air quality. Cadmium is an impurity in these metals; emission is not the result of the direct use of cadmium.

Unlike the phosphate industries, the secondary ferrous metals industry, i.e., iron and steel, is associated with "end-use" cadmium industries, primarily electroplating. Electroplated cars and trucks are reprocessed. Thus, in addition to direct regulation, the secondary metals industry will experience an economic impact from regulation of electroplaters.

The ferrous metal industry emits cadmium as an impurity in No. 2 scrap metal, primarily as a result of electroplating or galvanizing steel parts. Since the cadmium content of No. 2 scrap (consisting mainly of auto and truck bodies) is reported to be 45 ppm (mg/kkg) reprocessing can cause contamination.

Use of nonferrous substitutes will decrease the amount of available No. 2 scrap, currently valued at approximately $3,280 million, but a plating substitute should not affect availability.

We selected blast furnaces and steel mills as the most probable industry sector where cadmium impurity was a pollution factor. The 1977 value of this sector was reported as $41,998.2 million but this may have declined somewhat in the interim.[96] The majority of U.S. blast furnaces and steel mills are located either on the east coast or in the north central region. In 1977, the industry employed 441,900 workers of whom 350,000 were working in production. The Census of Manufacturers reported 396 companies with 504 plants of which 250 had more than 20 employees.[96]

IX. OTHER INDUSTRIES

Cadmium occurs or is used in a variety of other industrial materials such as coal, rubber goods, tires, motor oils, and fungicides. Because of its high neutron absorption, it was used in the control rods of nuclear reactors but has been replaced by boron and hafnium.

Coal contains 1 to 2 ppm cadmium and this industry is environmentally important because of the large quantity, about 500 million tons, consumed annually. In addition, lubricating, diesel, and heating oils contain cadmium but in amounts less than 0.5 ppm. Again, the quantity of use is critical.

Concentrations between 20 to 90 ppm in rubber tires result from impurities in zinc oxide. Cadmium oxide has been used to a minor extent to replace zinc oxide as an activator for curing rubber in some mechanical rubber goods, thus increasing the cadmium content.

X. OVERVIEW AND SUMMARY

Total domestic production of cadmium in 1978 was 1550 kkg; imports (metal and flue dust) totalled 3320 kkg, industry stocks were 2325 kkg, apparent consumption was 4469 kkg.[151] The trend in the U.S. since 1978 has been a continuing decrease in production coupled with a smaller increase in consumption.[154] Trends in cadmium use in each industry appears in Appendix V-A.

There appear to be some acceptable substitutes for noncritical cadmium uses in both the major use industry, electroplating (55%) and the minor use industry, alloys (4%). The remaining industries: pigments, plastic stabilizers, and Ni-Cd batteries would, in general, have to sacrifice quality, price, and performance, respectively.

Several factors seem to mitigate pollution in these industries. For example, because cadmium sulfide used by the pigment industry is relatively insoluble, the nickel-cadmium industry has an economic incentive to recover and recycle. Barium-cadmium stabilizers, because they tend not to leach from plastics, are minor contributors to cadmium pollution.

Finally, the use of substitutes in noncritical areas appears to be increasing. The phenomenon is a factor of price, rarity, and environmental concern.

Chapter 17

VALUE OF TECHNOLOGY — PHENOL

For which of you, intending to build a tower, sitteth not
down first, and counteth the cost, whether he have
sufficient to finish it?
Lest haply, after he hath laid the foundation, and
is not able to finish it, all that behold it
begin to mock him.
Saying, This man began to building, and was
not able to finish.

Jesus
Luke 14:28—30 (KJV)

Planning is important. As in knowing the cost of undertaking any endeavor, so must one determine the value of current technology to know what may be lost by imposing standards at various levels.

As in Chapter 16 for cadmium, we here assess the value of technology and various alternatives which would reduce phenol emissions. To avoid redundancy, the introductory remarks in that chapter apply here as well.

I. PHENOL-EMITTING INDUSTRIES

If you recall, we have discussed the various industries emitting phenolic compounds to waterways. These either manufacture phenol, use phenol in the production process, or emit phenolics as by-products.

Domestic production of phenol in 1978 was 2,756 million pounds; net exports were 228 million pounds; and production capacity was 3,390 million pounds. Total 1978 commercial value of production was $450 million and apparent consumption was 2,528 million pounds. In the past, production and demand grew 8 to 9%/year between 1969 and the peak growth year of 1974. Since then, demand has grown about 3% and production about 4%.[155]

Phenol and phenol-related industries are very dependent upon the petroleum and housing markets. For example, benzene prices fluctuate with oil prices. The price of phenol, in turn, is pushed up by cost and availability of the benzene feedstock. Sluggishness in the housing industry exerts a downward pressure on phenol prices, because of decreased demand for housing products made with phenol, such as fiberglass insulation, plywood, and carpeting. Poor economic performance by the auto industry also affects demand since, in attempting to increase auto appeal, intensive use has been made of plastics to reduce vehicle weight and improve fuel economy.

Most of the secondary products containing phenol are manufactured by mature industries with well-established markets, such as nylon and phenolic resin manufacturing. Future growth appears likely for the polycarbonate and epoxy resin industries (produced from bisphenol A) because of expanding applications to engineering, including protective coating, fiber reinforcement molding, and binding.

The top 13 industrial sources of phenolic pollution appear in Figure 24. The overall concentration scale used here covers five orders of magnitude. Most abatement techniques reduce concentration by two orders of magnitude. A reduction efficiency of 97% is typical. As you can see, the concentration from industrial effluent can span almost three orders of magnitude, as with the fiberglass insulation industry; but most ranges are within one order of magnitude. Industries are ranked from left to right according to their mean concentration levels.

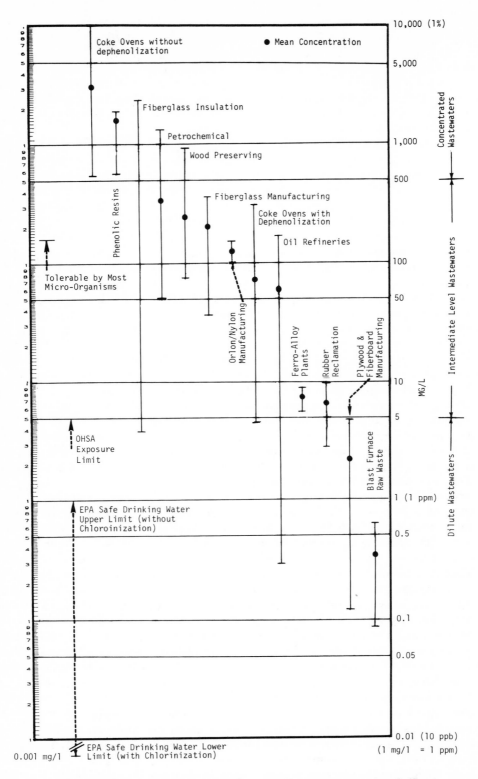

FIGURE 24.   Major phenol-emitting industries, ranked by mean concentration of effluent.

Three categories of wastewater concentrations are shown on the right side of Figure 24.[98] These categories were delineated and derived according to the type of treatment technology practical for each industry. Concentrated wastewaters have recovery potential but are, of course, the most ecologically detrimental. Intermediate level wastewaters are appropriate for biological processes, whereas dilute wastewaters must make use of physical-chemical methods. EPA safe drinking water limits based on odor provide the current standard reference level. These limits are in the lower range of the dilute wastewater category. The OSHA workplace exposure limit of 5 mg/ℓ is also shown.

There are five industries generating concentrated phenolic wastewater. They are coke ovens (coal and steel industry), phenolic resin manufacturing, fiberglass insulation, and the upper range in the petrochemical and wood preserving industries. Coke ovens are displayed twice in the figure, both with and without dephenolization, because about 59% of them do not employ the process. This is clearly a case where technology implementation could significantly reduce emissions. The effect on the coal industry would be to move its position from that of highest intensity phenol polluter to sixth highest intensity polluter, reducing phenolic mass emissions by more than 90%.

When recovery is employed, up to 98% efficiencies can be expected. In 1974 tight competition had effectively prevented economic recovery for coke plants, but now, with coal substitutes and fuel alternatives becoming more expensive, this may be a viable alternative.[98] Presently, recovery is practiced to some extent in the coke industry, with 4 of 61 plants utilizing phenol as a by-product. Compared with the older beehive process, the by-product recovery technique is the state of the art in coke production. In fact, 99% of the coke is currently produced utilizing this technique. Some 26 or more by-products, including phenol, are recovered from these operations.[156] Recovery volume for the industry has averaged about 30 million pounds.[155]

Even though a profit may not be generated, recovery can reduce abatement costs. It has greatest value where phenol is used in manufacturing, such as the fiberglass, phenolic resin, and petrochemical industries. These recovery processes, however, are inadequate to properly prepare concentrated effluents for discharge. A reduction through recovery of 98% will move the wastewater to the intermediate concentration level (5 to 500 mg/ℓ) but it may require further biological or physical-chemical treatment.

Wastewater recycling is a viable option for those industries using water for rinsing or nonchemical purposes, as in the manufacture of fiberglass and plywood. Recovery processes can be operated in tandem with this recycling if phenol concentration is large enough to make it economical. Recycling can be attractive because it eliminates or mitigates the pollution pathway and may reduce production costs.

The intermediate waste level is spanned by nine industries: the petrochemical and wood preserving industries, fiberglass insulation, fiberglass and orlon manufacturing, coke ovens with dephenolization, oil refineries, ferro-alloy plants, and rubber reclamation. The activated sludge process is the most frequently used method for industries with this intermediate level. Biological treatment processes affect all types of organic wastes and have very high reduction efficiencies into the 99% range. This can cover three orders of magnitude. Several physical-chemical methods are also available for treating and reducing intermediate level wastewater to the dilute level.

Wastewater from the plywood manufacturing industry and blast furnaces are in the dilute range, but reduction is still possible. In general, reduction efficiencies are very low for biological techniques, ranging from 20 to 70%. However, use of an ozone treatment process, which can reportedly achieve a 98% reduction would reduce phenol concentration from 160 to 3 ppb (0.003 mg/ℓ).[98] To reduce concentrations from a very high level, as for coke ovens, to the dilute range and beyond requires tertiary treatment at high cost. However, concentrations generally are not high and most industries would probably require only secondary treatment to reduce residues to relatively low levels.

Table 22
SEVEN SIGNIFICANT PHENOL-EMITTING INDUSTRIES WITH EMPLOYMENT AND PRODUCTION VALUE

Industry	Number of plants	Employment (thousands)		Value ($ million)	Sources
		Total	Production		
Coke ovens	63	30.6	26.2	873	COM[a] 1972
Phenol manufacturing	14	3.4	2.2	934	COM[a] 1972 (derived)
Phenolic resins	41	7.0	4.5	570	COM 1972 (derived)
Petrochemical	453	135.8	87.5	10,696	COM 1972
Blast furnaces	364	469.1	379.3	23,947	COM 1972
Plywood	477	68.5	61.0	5,177	COM 1977
Oil refineries	349	102.5	70.8	91,689	COM 1977

[a] Census of Manufacturers[96]

Table 22 (derived from Census of Manufacturers data[96]) shows the employment and production values for the 7 emitting industries we have selected. Coke ovens, phenolic resin and phenol manufacturing, oil refineries, and the petrochemical industry are high concentration emitters. The plywood industry and blast furnaces, although emitting low concentrations, are economically important. Manufacturing plants, for the most part, are located near feedstock sources and industrial regions requiring phenol for resin and plastics production.

Blast furnace location is determined primarily by the location of other heavy industry such as that found in the north central region of the country. Coke ovens are generally used in conjunction with blast furnaces, and their location and density correlate roughly with those of blast furnaces. Petrochemical plants are also very roughly correlated with the location of phenol manufacturing plants (a subset of the petrochemical industry). Plywood plants are mainly located on the west coast and in the deep South. The economic impacts on the states or regions where these plants are located would be considered in any benefit assessment for standard setting.

II. SUBSTITUTES

Almost all phenol-related products may be replaced to some degree. There are, however, some products so unique in their physical characteristics that there are few or no substitutes. Most nylon products fit this description.

Some industries, such as the plywood industry, have begun utilizing substitutes and extenders that become more competitive with rising cost of pollution abatement. Appendix V-B describes these products and indicates the extent to which their use could reduce that of phenol. Table 21 (presented in Chapter 16) can serve as a general guideline.

Many phenol products are preferred because of price and physical characteristics. For example, substitutes for phenolic resins and for bisphenol A, used to make both epoxy (phenolic) and polycarbonate resins, are available only at higher cost and/or substantial loss in quality because of certain deficient physical properties. Over the past 5 years, demand growth for phenolic resins has slowed to 2% from a high of 7% and that for bisphenol A to 8% from a 15% level.

Uses not discussed are varied and comprise 18% of all phenol consumption. If we assume 50% of the phenol so used can be replaced, an additional 9% reduction will occur. Total usage can be reduced about 47% or, more realistically, 40% to 50% by the use of substitutes.

Table 23
PRODUCTION OF PLASTICS AND
RESINS

Division	Percentage of total production in 1967
Cellulosics	3.5
Vinyls	19.5
Styrenes	16.0
Polyolefins	30.5
Acrylics	2.0
Polyesters and alkyds	7.5
Urea and melamines	4.5
Phenolics	7.5
Miscellaneous resins	9.0
Total	100.0

III. PLASTICS AND RESINS

Phenol is only one of many feedstock sources used in plastics and resins production, as depicted by Table 23 (from U.S.D.I. data). Between 1962 and 1967, use of phenolics for this purpose increased 52%. Still, in 1967, it represented only 7.5% of the total, while vinyls, styrenes, and polyolefins constituted the remaining 2/3. Although phenol serves specific purposes, in a broad value of techology sense, it plays but a small role in an enormous industry.

Chapter 18

ECONOMIC IMPACT AND COSTS OF WATER TREATMENT

Never ask of money spent
Where the spender thinks it went
Nobody was ever meant
To remember or invent
What he did with every cent

Robert Frost
The Hardship of Accounting (1930)

Perhaps we can do a bit better in planning for the expense of regulation. Here we discuss methods for weighing monetary cost and their application to cadmium and phenol water pollution.

I. ECONOMIC IMPACT

In general, pollution abatement most adversely affects two types of industries:

1. Those with a high total mass efflux of pollutant and a low intensity or concentration
2. Those with a low mass efflux and a high intensity

Of course, most industries would be placed somewhere between these limits. The high mass efflux/low intensity situation requires that production be reduced. This can be achieved by switching to substitute products with less toxic pollution, lowering demand through regulation or taxation, or by including social costs in the total product cost. The reduction effect on demand depends upon price elasticity. If prices are inelastic, as they are over the short run in the petroleum industries, a surtax may not significantly effect demand.

The low mass efflux/high intensity case may require developing and adopting new technologies. The impact of abatement is not as severe in this case, because it is relatively easier to reduce intensity than mass efflux. Resulting economic impacts will be higher costs and a reduction in demand.

Most firms should not be drastically affected by abatement procedures, although in every industry marginal firms would likely be forced out. The economic effect on the consumer is reduced demand at a higher price, and on industry, reduced competition. The bottom line for the consumer is reduced purchasing power (inflation). However, the consumer may not, in sum, be worse off because of intangible benefits in higher environmental quality. This is, of course, a matter of degree and of perception. These consumer impacts are difficult to assess for a single industry or for the abatement of a single pollutant.

The impact on industry is more readily derived because the firms are known and the government tabulates certain industrial performance data. Several industrial problems may arise from abatement. For example, the rate of capital obsolescence may be increased by new technology abatement equipment, as it would in the steel industry. There may also be a substantial influence on foreign trade resulting from reduced market competitiveness of U.S. companies because of their costly goods. Conversely, domestic markets may become more susceptible to foreign penetration.

In terms of capital turnover and investment, abatement requirements remove funds available for other investments. Since intangible pollution abatement benefits to society are not considered by the marketplace, the effect on reduced demand and investment is reduced productivity. The ultimate effect in capital markets is a lower demand for stocks and bonds in the affected industry.

On the larger (macroeconomic) scale, inflation can pressure the Federal government to use fiscal policy as a stabilizer. This can create a paradox for government, because one major policy objective of most administrations is the achievement of high employment. Abatement of pollution can produce both unemployed workers from closed marginal plants and reduced productivity in other firms. The public may not want to trade environmental quality for high unemployment. On top of these pressures, government must increase expenditures in tandem with the private sector to fund its programs and support regulatory agencies. This requires taxes and increased debt with additional economic impact.

Even interest rates are affected by resulting economic stresses. These in turn can affect housing starts. In fact, the effects of pollution abatement are believed to have reduced housing starts in 1975 by some 90,000 units.[156] We are all aware that other factors have since further depressed the housing industry.

II. COST EVALUATION METHODS

Two techniques are used in evaluating cost. The first is the *crude annual cost* approach which does not consider variation of the value of money with time. A formula to derive the crude annual cost using this approach is presented below. For the sake of simplicity, capital cost of equipment is apportioned evenly over the lifetime of the facility and the annual operating cost is added to this value.

$$\text{Crude Annual Cost} = \frac{\text{Capital Costs}}{\text{Operating Lifetime}} + \text{Annual Operating Cost} \qquad (9)$$

Although this approach does not conform with usual accounting or economic evaluation techniques, it gives a cost intermediate among results of other techniques and is easier to understand.

The second technique, normally used in cost/benefit analyses, is known as the present worth, or present value, method. This approach considers the total value of money which may be set aside to later conduct an activity with future costs. The present worth of a series of uniform payments A at n fixed intervals in the future is defined for a discount rate, i, as:

$$\text{Present Worth} = A\,\frac{(lti)^n - 1}{i(lti)^n} = \text{PWF}_{i,n} \qquad (10)$$

The value of the factor is defined by the values of n and i and is termed the *annuity present worth factor* (PWF). In this analysis, the annuity present worth factor is applied to the future annual operating costs and initial capital cost is added to obtain the present worth:

$$\text{Present Worth} = \text{Capital Cost} + (\text{PWF}_{i,n}) \times \text{Annual Operating Cost} \qquad (11)$$

The Office of Management and Budget (OMB) suggests a 10% discount rate for evaluating capital investment. According to the OMB, "the prescribed discount rate of 10 percent represents an estimate of the average rate of return on private investment, before taxes and after inflation."[158] Of course, capital costs for different firms and industries vary according to their financial structure.

Inflation complicates an evaluation of time-dependent costs. Available cost information is cited from different points in time. Rather than adjust values for inflation, we have tabulated all cost figures with the corresponding year. We assume the 10% discount rate will account for time value and inflation.

We further assume the average life of equipment to be 20 years, again a representative figure for pollution abatement assessment.[158] Since we will make all cost evaluations using the same life and discount rate, the annuity present worth factor is 8.5136. For different equipment lives and discount rates, the present worth factor is computed from:

$$PWF_{i,n} = \frac{(1 + i)^n - 1}{i(1 + i)^n} \qquad (12)$$

We must make several economic assumptions before evaluating control costs. For example, waste concentration level affects cost of treatment. It is common sense that treatment of a high influent concentration may cost more than a moderate concentration, but we must also be aware that reducing an extremely low influent concentration may also be expensive because quality and operational controls are difficult to apply.

Economies of scale affect treatment systems with different capacities. A low waste volume, usually measured in million gallons per day (MGD) has greater processing costs per gallon than a large volume system. Treatment costs are usually represented on a per gallon basis to indicate relative cost and capital costs are expressed as dollars per thousand gallons per day, since capital investment essentially purchases capacity. Similarly, annual operating cost is expressed as dollars per thousand gallons treated, since operating costs purchase volume.

We assumed cost information epitomizes costs of a typical treatment facility. We then re-expressed relative costs in terms of total capital and operating costs per plant, applying the two cost evaluation techniques. Cost figures were tabulated according to marginal reduction of effluent concentration for each waste treatment stage.

The typical treatment stage reduces concentration 98% or greater. However, water with a concentration of 10,000 mg/ℓ must be put through 2 treatment stages for a 2 orders of magnitude reduction to 4 mg/ℓ (99.96% cumulative reduction). If the reduction (r) for each stage is known, cumulative reduction (cum) can be easily obtained by adding reductions at each stage.

$$r_{cum} = r_1 + r_2 - r_1 r_2 \qquad (13)$$

A. Application to Cadmium

Three cost evaluations for waste treatment are shown in Tables 24 and 25. These tables combine data on concentrations,[66] costs derived from several EPA sources, and reduction of effluent.[65,98] The first table describes the three categories of the electroplating industry. Note that job and captive shops must use chromium and cyanide oxidation prior to efficiently using a cadmium treatment process. This is an additional cost other cadmium-related industries do not have. Thus, a job shop incurs costs of $1.5 million in present worth before obtaining any cadmium reduction.

The first stage involves pH adjustment, the most significant reduction stage. The next step is typically clarification, which involves filtration and sedimentation and reduces concentration an additional 4%. Further steps have only marginal effect in reducing concentration.

Table 25 epitomizes treatment costs for a typical pigment plant and a typical mining and smelting operation. Pigment production results in small-size particles usually less than 1 μm in diameter, which because they are generally insoluble, may facilitate removal from wastewater. In fact, approximately 99.5% of the influent cadmium from a pigment plant is in an insoluble form (200 mg/ℓ).

Table 24
ELECTROPLATING

Predominant processes	Water usage volume (GPD)	Reduction factor	Concentration (mg/ℓ)	Capital costs ($)	Annual costs ($)	Present worth ($ million)	Crude annual costs
Job shops	38,200						
Cr reduction[a]		—	26	—	—	—	—
Cyanide oxidation[a]		—	26	137,900	52,440	0.58	0.059
pH Adjustment		—	26	301,390	140,440	1.50	0.156
Clarification		14	1.82	313,910	145,880	1.55	0.162
Reverse osmosis		33	0.78	350,200	158,580	1.70	0.176
		100	0.26	391,930	189,970	2.01	0.210
PB Manufacturing	18,800		26				
Cyanide oxidation[a]		—	26	—	—	—	—
pH Adjustment		—	26	65,546	21,342	0.25	0.025
Clarification		14	1.82	74,476	25,717	0.29	0.029
Reverse osmosis		33	0.78	91,264	30,717	0.35	0.035
		100	0.26	118,803	46,167	0.51	0.052
Captive shops	277,000		26				
Cr reduction[a]		—	26	—	—	—	—
Cyanide oxidation		—	26	167,100	59,208	0.67	0.068
pH Adjustment		14	1.82	1,132,560	373,668	4.31	0.430
Clarification		33	0.78	1,223,350	399,968	4.63	0.461
Reverse osmosis		100	0.26	1,470,710	473,650	5.50	0.547
				2,062,710	851,270	7.45	0.954

[a] Efficiencies dependent upon cyanide destruction and chromium reduction.

Table 25
PIGMENTS AND MINING AND SMELTING

Treatment process	Reduction factor	Concentration (mg/ℓ)	Capital costs ($)	Annual costs ($)	Present worth ($)	Crude annual cost ($)
Pigments[a]						
No control	0	200	0	0	0	0
Settling and pH adjustment	571	0.35	75,000	35,000	373,000	39,000
Clarification	1,000	0.2	85,000	37,500	404,000	42,000
Holding tanks to attenuate hourly variation	2,000	0.1	235,000	49,500	656,500	62,000
Mining and smelting[b]						
No control	0	15	0	0	0	0
Liming to pH 10.5 and settling	30	0.5	0.25	0.076	0.899	0.089
Wastewater volume reduction (35%)	3[c]	0.5	0.176	0.075	0.814	0.084

Note: Pigment costs expressed in dollars; mining and smelting cost in millions of dollars.

a   Waste volume: 24,000 gal/day; cost data: 1980.
b   Waste volume: 0.369 MGD; cost data: 1971.
c   After mixing.

Note in Table 25 that 99.93%, reduction factor of 571, is initially removed through settling of insoluble pigment and pH adjustment of the remaining ion solution waste water. Further reduction, however, contributes only minutely to the cumulative reduction.

Mining and smelting plants have a mean raw waste water concentration of about 15 mg/ℓ.[65,98] Treatment is accomplished by liming to a pH of 10.5 and settling to remove metals, such as arsenic, mercury, selenium, zinc, and cadmium. This process typically has a reduction factor of 30, or 97%. While it does not affect effluent concentration, it does affect downstream concentrations by increasing stream dilution.

B. Application to Phenol

Two cost evaluations for phenol waste treatment, after Patterson, are shown in Tables 26 (a hypothetical coke plant) and 27 (a hypothetical petroleum refinery).[98] The first represents the cost of treating intermediate to concentrated wastewaters with approximately 500 mg/ℓ influent concentration. Data were obtained for a coke plant with a 0.5 MGD waste volume. Results are tabulated according to marginal reduction in concentration but the numbers represent cumulative effect. For a high influent concentration, typical for coke ovens and most phenol-emitting industries (see Figure 24), the first stage in treatment usually involves a recovery process. These processes are generically referred to as dephenolization. The first stage reduces concentration by 98.6%; 98 to 99% is typical for all phenol treatment processes. This translates to a reduction factor of 72.

During the second stage, lime and steam are added and concentration reduced by another 80%, a cumulative reduction factor of 360. Marginal costs for the second stage are small, increasing present worth from $11.8 to $13.25 million. The marginal reduction in concentration, however, was also quite small. During the third stage the activated sludge process is applied. This is a typical biological treatment process used by industries and municipalities. The marginal cost and concentration reduction here are both significant. The entire treatment system reduces influent concentration by a factor of 36,000 (99.99%) and still yields an effluent concentration of 0.010 mg/ℓ or 10 ppb.

Table 27 epitomizes the costs for treating dilute to intermediate wastewaters (approximately 5 mg/ℓ influent concentration). The data are for a petroleum refinery with a 2 MGD waste volume, 4 times the capacity of the coke oven example. There are two treatment stages. First, the activated sludge process is applied to the 6.63 mg/ℓ influent wastewater, just as it was to the 1 mg/ℓ wastewater in the coke oven example. Then, the activated carbon process is applied. The reduction factor is 5,100 (99.98%) and the effluent concentration is 0.0013 mg/ℓ or about 1 ppb.

Table 26
COKE PLANT (2,660 TON/DAY)

Treatment process	Reduction factor	Concentration (mg/ℓ)	Capital costs ($ million)	Annual costs ($ million)	Present worth ($ million)	Crude annual cost ($ million)
No control	0	360	0	0	0	0
Ammonia stripping and dephenolization	72	5	4.48	0.86	11.80	1.08
Lime and steam addition	360	1	4.65	1.00	13.25	1.23
Activated sludge	36,000	0.010	11.93	1.15	21.75	1.75

Note: Waste volume: 0.5 MGD; cost data: 1974.

Table 27
PETROLEUM INDUSTRY

Treatment process	Reduction factor	Concentration (mg/ℓ)	Capital costs ($ million)	Annual costs ($ million)	Present worth ($ million)	Crude annual cost ($ million)
No control	0	6.63	0	0	0	0
Activated sludge	50	0.133	0.73	2.0	17.26	2.037
Activated carbon	5100	0.0013	2.2	6.0	53.28	6.110

Note: Waste volume: 2 MGD; cost data: 1974.

Chapter 19

RISK TO HUMANS AND WILDLIFE FROM CADMIUM
(FOR MODEL CALCULATIONS)

*The health of the people is really the foundation upon
which all their happiness and all their powers
as a state depend.*

Benjamin Disraeli
Speech (July 24, 1877)

This section correlates the chronic and acute dose/effect relationships from Chapter 13 and both ambient and effluent levels from each industry considered as derived via our model in Chapter 11.

I. RISK TO AQUATIC SPECIES

As previously described, a wide range of chronic effects in both vertebrate and invertebrate aquatic species may occur as a result of exposure to either cadmium or phenol. In Chapter 13, we arbitrarily chose effects from the published literature to illustrate the method.

Table 28 compares the range of selected observed effects with that of ambient and effluent concentrations from each industry. We have used effluent rather than attenuated (settling) range to give a worst case situation. However, assuming as we have, that 1 day is required for complete input of contaminated water to municipal water holding areas, the number of industries exceeding effect levels does not change whether considering attenuated or non-attenuated concentrations. The acute and chronic levels selected may be exceeded for both finfish and shellfish by source waters of all nine investigated industries. The "small river" concentration presents a more plausible worst case situation. In this instance, five industries may exceed the lowest selected effect level for invertebrates and three the level for finfish. Some industries along "average" size rivers may exceed lowest effect levels (Table 29). Note that ambient concentrations at the higher end of their range may also exceed lowest effect levels.

Effluent from certain industries may contain other compounds more toxic to aquatic life than that from other industries. An example is the cyanide and sulfate solutions from electroplating which are more hazardous to aquatic species than the insoluble sulfide from zinc ore tailings. We have not estimated these differential effects.

II. HUMAN HEALTH RISKS

We have estimated health risks from both ambient levels and concentrations which may result from industrial waste water emissions.

A. Risk from Ambient Exposure via Water Pathways

Health risk from ambient exposure in water pathways are derived from dose-effect estimates and the populations at risk.

Cadmium in water supplies is less than 0.4 ppb and raw water averages about 5.1 ppb (Table 8) with an upper range of 500 ppb. Cancer from ingestion of cadmium in water at 0.4 ppb, if based upon a linear extrapolation to the U.S. population from Figure 16, would result in a high estimate of about 25 to 0.25 manifestations of cancer per year as a midrange

Table 28
RANGE OF AMBIENT AND EFFLUENT CADMIUM LEVELS COMPARED WITH ADVERSE EFFECT LEVEL RANGES

Raw water samples	Vertebrates		Invertebrates	
	Acute effects (LC$_{50}$) (10 ppb to 10 ppm)	Chronic effects (3 ppb to 10 ppm)	Acute effects (LC$_{50}$)	Chronic effects (1 ppb to 2.5 ppm)
Effluent — source (9 industries) 0.02 to 16 ppm[a]	All industries exceed lowest effect level	All industries exceed lowest effect level	Not calculated	All industries exceed lowest effect level
Effluent — diluted small river normal flow (9 industries) 0.1 to 1172 ppb[a]	3 Industries exceed lowest effect level	4 Industries exceed lowest effect el	Not calculated	5 Industries exceed lowest effect level
Effluent — diluted average river normal flow (9 industries) 0.01 to 9.25 ppb[a]	None of the industries exceed lowest effect level	2 Industries exceed lowest effect level	Not calculated	3 Industries exceed lowest effect level
Ambient — 0.0 to 28.0 ppb	The higher end of the ambient range exceeds lowest effect level	The higher end of the ambient range exceeds lowest effect level	Not calculated	The higher end of the ambient range exceeds lowest effect level

[a] The three types of electroplating operations are each considered here to be separate.

Table 29
CADMIUM-EMITTING INDUSTRIES EXCEEDING LOWEST EFFECT LEVELS FOR AQUATIC SPECIES

Exceeds lowest effect level

	Acute (LC$_{50}$)				Chronic			
	Vertebrates		Invertebrates		Vertebrates		Invertebrates	
	River size				River size			
Industry	Small	Average	Small	Average	Small	Average	Small	Average
Mining/smelting	X							
Electroplating					X	X	X	X
Job shops					X		X	
PB manufacture							X	
Captive shops	X							
Pigments					X	X	X	X
Plastics					X			
Ni-Cd batteries								
Iron and steel								
Phosphate fertilizers	X				X	X	X	

CRITERIA FOR RISK TO WILDLIFE

There are two risk criteria for effects to aquatic species, and wildlife in general, which must be considered when conducting a risk assessment. These are

- Do significant population reductions result from the levels of pollution discussed?
- Are endangered or threatened species affected?

Significant population reduction is a value judgment. We have used the LC_{50} or lethal concentration adequate to kill 50% of a particular aquatic species, as a significant reduction. We recognize that this permits a relatively high level of pollution and includes all species regardless of importance to commercial or recreational fishing or to the food chain. It requires a more in-depth assessment to support a case for significant population reduction for acute effects than for chronic effects, although sub-lethal chronic effects on behavior and physiology can have serious impact.

Of course, in actual standard setting all these effects would be considered. The Water Program of EPA, in preparing criteria documents uses effects on sensitive species to determine the amount which would be protective of 95% of the aquatic species likely to be present. Any criterion requires at least eight valid acute and valid chronic studies on sensitivity of the species in question to the pollutant.[160]

We did not prepare a complete assessment of effects on endangered or threatened aquatic species by either cadmium or phenol. In general, the cadmium emitting industries appear to be located in highly industrialized areas where the general population level may be presumed adequate to have long since disposed of sensitive aquatic populations. Cadmium's persistence, however, might require an assessment as to whether it reaches isolated localized aquatic habitats. Persistence may also conceivably make cadmium a threat via the food chain to certain endangered and threatened land species.

---

estimate, and zero at lower levels (including, of course, the no observable effect level). Since the data upon which we base these effects may be unreliable, and epidemiology has not suggested cause/effect at the high level, the midrange estimate appears to be more meaningful. There are no acute or noncarcinogenic chronic effects from cadmium in drinking water. Intake of raw water at the maximum level (Table 8), if continuous, could result in some slight level of kidney damage (Figure 15). Since this level of intake is not anticipated, we will leave both acute and chronic noncarcinogenic effects at zero for ambient levels in water.

Cadmium, if carcinogenic, might cause a few tumors at the high range estimate. That estimate, however, is difficult to substantiate.

B. Risks from Industrial Water Emissions

We will now derive impacts as a result of exposure to concentrations expected from industrial sources as we did for those resulting from ambient levels, from dose-effect estimates, and from estimates of population at risk. The concentrations and the estimates of population at risk are from the exposure model for each industry (Tables 12 and 13 in Chapter 11). Cancer estimates are based upon a linear dose-effect relationship, assuming equivalent risk from high exposure to a small population on a small river equal to low exposure to a larger population on average or large rivers. Thus, we had only to use the average size river to estimate the cancer risk.

Noncarcinogenic chronic exposure estimates are for both small rivers at high concentration and average size rivers at low concentrations. Note that values in the tables are only for use in this analysis. Again, they should not be cited for other purposes since our model has been neither verified nor validated.

Table 30 summarizes cadmium health impact estimates. The first three main columns are

Table 30

CADMIUM HEALTH RISK ESTIMATES FROM INDUSTRY WATER PATHWAYS[a]

Industry	Industrial concentrations		Population × 10^6		Cancers			Chronic effects	
	Small rivers	Average rivers	Small rivers	Average or larger	Maximum	Mid-range	Minimum	Small rivers	Average or larger
Mining/smelting	1031 ppb	8 ppb	.25	2.5	7	<<1	0	125 × 10^3 k/d[b]	n/e[c]
Electroplating									
Job shops	8.5	0.06	10	100	2	<<1	0	n/e	n/e
Pb manufacture	0.06	0.004	10	100	<1	<<1	0	n/e	n/e
Captive shops	583	4.5	10	100	100	1.0	2 × 100	2 × 10^6 k/d	n/e
Pigments	0.2	0.002	0.22	2.2	<<1	<<1	0	n/e	n/e
Plastics	0.1	0.0007	7	70	<<1	<<1	0	n/e	n/e
Ni-Cd batteries	1.4	0.01	0.36	3.6	<<1	<<1	0	n/e	n/e
Iron and steel	0.37	0.003	4.9	49	<<1	<<1	0	n/e	n/e
Phosphate fertilizers	306.1	2.3	3.3	33	30	<<1	0	330 × 10^3 k/d	n/e

[a] These values are only for use in the analysis and should not be cited for other purposes.

[b] k/d = Kidney damage.

[c] n/e = No effect.

the industries, concentrations, and populations which were shown in Table 12. Concentrations are converted to cancer risk via Figure 16 for worst case, mid-range, and zero effect levels, and these risks are then multiplied by the population at risk to derive total cancers. We obtained noncancer chronic risks in the same manner, except that Figure 15 is used for the conversion and a smaller population for small river cases.

Cancer health effects from cadmium are very low unless one assumes the worst case for mining and smelting and for some aspects of the electroplating industry. Chronic effects in the form of kidney disease may occur among those living along small streams, but not for others.

As a worst case, we estimate about 2.5 million cases of kidney disease from mining and smelting, electroplating, and phosphate fertilizer use on small streams. These effects are drawn from those observed in the Itai-Itai disease outbreaks in Japan. This represented an extreme situation because those affected had a worst case exposure from decades of local zinc smelting contamination of rice and fishes which constituted the major part of the diet.

There were 13 million reported cases of kidney disease in the U.S. as of May 1979, according to the National Kidney Foundation.[161] However, several factors may mask the incidence of this disease. For one thing, diagnoses probably lag behind incidence. In addition, people may die of other causes before the disease is fully manifested. Because of this, the kidney damage estimate arrived at here, that waterborne cadmium might be a contributing factor in, at most, about 20% of all kidney disease does not appear unreasonable.

C. Cumulative Noncarcinogenic Chronic Risk

In addition to industrial water sources, there is cumulative exposure to cadmium and phenol via other routes, i.e., air and food pathways. Moreover, as previously described, for special populations other pathways are important, as in the case of smokers for cadmium. These factors are considered here.

Small river flow data is again used for worst case estimates. We adjusted estimates of exposure from industry waste water sources plus ambient levels from water, air, food, and smoking (as a special case) for equivalent intake and then tallied them to find the cumulative chronic exposure (Table 31). The number of people exposed to these levels were converted into health impact via the relationship in Figure 15. Air and smoking values of 1 ppb and 8 ppb, respectively, are then multiplied by 5 to account for the fact that 30% of the inhaled cadmium is retained compared with only 6% of that ingested (see Chapter 12). This provides an equivalency for inhalation vs. ingestion pathways. Inhalation values for cigarette smoking are 4 mg per pack (approximately 4 ppb per pack). About 54 million people in the U.S. smoke between 1 and 2 packs per day.[162] This is approximately 25% of the total population.

Kidney disease is anticipated in 5.6 million smokers and 1.6 million nonsmokers; a total of 7.2 cases or approximately one half the amount detected each year by the Kidney Foundation. Therefore, in this worst case, about one half of these cases would be related to cadmium exposure from all sources.

Cadmium may be a "moderate" health risk in water. It may, however, be a more serious hazard considering all sources cumulatively. Evidence in the literature is inadequate to confirm or disprove these risk estimates.

Table 31
CADMIUM HEALTH RISKS FROM ALL SOURCES

Industry	Industrial sources Small rivers	Industrial sources Average rivers	Population × 10^6 Small rivers	Population × 10^6 Average rivers	Ambient Water 0.4/5 ppb[d]	Ambient Air[b] 1 ppb × 5	Ambient Food 75 ppb[c]	Subtotal Small rivers	Subtotal Average rivers	+2 Packs/day smokers 8 ppb × 5	Total ppb Small rivers	Total ppb Average rivers	Percent attributed to water Non-smokers Small rivers	Percent attributed to water Smokers Average Rivers	Chronic Effects Nonsmokers Small rivers	Chronic Effects Nonsmokers Average rivers	Chronic Effects Smokers (25% total pop.) Small rivers	Chronic Effects Smokers (25% total pop.) Average rivers
Mining/smelting	1031 ppb	8 ppb	0.25	2.5	5 ppb	5 ppb	75 ppb	1011	88	40 ppb	1141	128	90 (90)	9 (6)	50% (90 × 10^3)	n/e	55% (34 × 10^3)	n/e
Electroplating																		
Job shops	8.5	0.06	10	100				88.5	80.1		128.5	120	10 (7)	+	n/e	n/e	n/e	n/e
Pb maintenance	0.06	0.004	10	100				80.1	80.0		120	120	+	+	n/e	n/e	n/e	n/e
Captive shops	583	4.5	10	100				663	84.5		703	124.5	88 (83)	5 (4)	40% (5 × 10^6)	n/n	45% (1.1 × 10^6)	n/e
Pigments	0.2	0.002	0.22	2.2				80.2	80		120	120	0.25 (0.16)	+	n/e	n/e	n/e	n/e
Plastics	0.1	0.0007	7	70				80.1	80.1		120	120	+	+	n/e	n/e	n/e	n/e
Ni-Cd batteries	1.4	0.01	0.36	3.6				81.4	80		121	120	2 (1)	+	n/e	n/e	n/e	n/e
Iron and steel	0.37	0.003	4.9	49				80.4	80		120.4	120	0.5 (0.3)	+	n/e	n/e	n/e	n/e
Phosphate fertilizer	306.1	2.3	3.3	33				386.1	82.3		426	122	79 (72)	3 (2)	30% (5 × 10^5)	n/e	35% (2 × 10^5)	n/e

+ = <0.2%

Note: n/e = No effect based upon entry D on Figure

a These values are only for use in the analysis and should not be cited for other purposes.

b Corrected for greater health impact — air = 30% retained vs. 6% for food per person per day.

c May reach 300 ppb for 15% of population.

d Drinking water: 4 ppb; raw water: 5 ppb.

Chapter 20

RISK TO HUMANS AND WILDLIFE — PHENOL

Salus populi suprema lex.
The people's safety is the highest law.

Legal and political maxim

We will proceed in determining phenol risk to humans and aquatic species in the same manner as for cadmium.

I. RISK TO AQUATIC SPECIES

Table 32 lists phenol emitting industries which may exceed effect levels for aquatic species. As with cadmium, the acute effects are based on LC_{50} (see Chapter 19 for discussion).

One should also be aware of certain factors when applying these acute and chronic effect levels to exposure to determine risk. For example, at the lower end of the chronic effect range some subtle effects, such as reduced growth, are difficult to quantify in terms of population survival. Effects on aquatic plants, unquantified here, can impact, via food chains and webs, organisms at higher trophic levels. In addition, manifestation of both chronic and acute phenol effects is dependent on the species affected as well as a number of water conditions.

As discussed, in the case of cadmium, attentuation is due to settling or sedimentation — for phenol, the action of bacteria. As with cadmium, 1-day phenol attentuation reduces concentration so slightly that, with one exception, the number of industries exceeding lower effect levels remains constant for attenuated and nonattentuated concentrations. Table 33 presents the range of ambient and effluent concentrations and subsequent effect on aquatic species.

Unlike cadmium, ambient phenol levels apparently do not adversely affect aquatic organisms. The undiluted effluent of most of the investigated industries exceeds acute and chronic effect levels as do a number of industries if located on "small" rivers. Those along "average" sized rivers, under normal flow conditions, exceed only the chronic effect level. Significant population reduction appears unlikely for any investigated industry under this scenario.

II. HUMAN HEALTH RISKS

These are determined in the same manner as for cadmium.

A. Risk from Ambient Exposure via Water Pathways

Assuming a worst case of 10 million people exposed to 1.4 $\mu g/\ell$ of phenol through community water supplies (Figure 9), the cancer risk to an individual would range from about 2×10^{-7} cancers per year (Figure 16) at the maximum level, and zero if the substance is not carcinogenic. The worst case estimate would be about 0.2 cancers per year in the total population. Intake from sources other than drinking, i.e., finfish and immersion, could not exceed this value. No chronic or acute effects would occur.

There are neither chronic nor acute effects for phenol or cadmium from ambient water levels. Cancer from phenol, if indeed it is carcinogenic, results in less than one tumor per year in the affected populations.

B. Risk from Industrial Water Emissions

Table 34 summarizes our phenol health impact estimates. The first three major headings

Table 32

PHENOL-EMITTING INDUSTRIES EXCEEDING LOWEST EFFECT LEVELS FOR AQUATIC SPECIES

	Exceeds lowest effect level								
	Acute (LC_{50})						Chronic		
	Vertebrate			Invertebrate			Vertebrate		
		Dilution based on river size			Dilution based on river size			Dilution based on river size	
Industry	Source	Small	Average	Source	Small	Average	Source	Small	Average
Coke ovens									
No dephenolization	X	X		X		X	X	X	
Dephenolization	X			X			X	X	
Phenol manufacturing	X	X		X	X	X	X	X	
Petrochemical	X	X		X	X	X	X	X	
Oil refineries	X			X		X	X	X	
Plywood						X			
Blast furnaces						X			
Phenolic resins	X	X		X	X	X	X	X	

Table 33

RANGE OF AMBIENT AND NONATTENUATED EFFLUENT PHENOL LEVELS COMPARED WITH ADVERSE EFFECT LEVEL RANGES

Phenol effects on aquatic species

	Vertebrates		Invertebrates	
Raw water levels	Acute effects (LC$_{50}$) 9.4 to 63 ppm	Chronic effects 0.1 to 10 ppm	Acute effects (LC$_{50}$) 14 to 780 ppm	Chronic effects
Effluent — source 0.34 to 3,360 ppm	6 Industries[a] exceed lowest effect level	All industries[a] exceed lowest effect level	6 Industries[a] exceed lowest effect level	Not calculated
Effluent — diluted small river normal flow 0.001 to 33 ppm	4 Industries[a] exceed lowest effect level	6 Industries[a] exceed lowest effect level	3 Industries[a] exceed lowest effect level	Not calculated
Effluent — diluted average river normal flow 0.01 to 260 ppb	None of the industries exceed lowest effect level	3 Industries[a] exceed lowest effect level	None of the industries exceed lowest effect level	Not calculated
Ambient 0.0 to 28.0 ppb	No effect	No effect	No effect	Not calculated

[a]   Coke plants with and without dephenolization each considered here as separate industries.

ENDANGERED SPECIES

   Most phenol emitting industries, such as coke plants, chemical, fiberglass, and plastic manufacturers are located in metropolitan areas along bodies of water. The wood preserving and plywood industries, however, are typically located in rural areas because lumber is their major resource input. The largest plywood producing states are North Carolina for hardwood and Oregon for softwood; predominantly rural states where waterways are, in general, less contaminated than those in urban areas, and are possibly more likely habitats for endangered species. An assessment of effects on endangered and threatened species would require consideration of the location of plants, potentially affected populations, and the attenuation rate of phenol.

   Effulent from the plywood industry does not exceed acute effect levels. However, even the slightest adverse impact, as with certain lower level chronic effects, could conceivably threaten existence of some of these species.

---

are the industries, concentrations, and populations shown in Table 12. The concentrations are converted to cancer risk via Figure 23 for maximum, mid-range, and zero effects and, as with cadmium, these risks are then multiplied by the population at risk to derive total cancers for each case. The mid-range risk for cancer is assumed to be two orders of magnitude lower than the range (EPA) in Figure 23. This is consistent with the cadmium mid-range estimates. We obtained chronic risks in the same manner as for cadmium except that we used Figure 22 for the conversion and a smaller population in the small river case.

   Cancer health effects from phenol are low except for worst case estimates for some coke ovens, petrochemical plants, and oil refineries. There are no chronic effects from water alone.

Table 34
PHENOL HEALTH ESTIMATES FROM INDUSTRY WATER PATHWAYS[a]

Industry	Industrial sources		Population × 10⁶		Cancers			Chronic effects	
	Small rivers (ppm)	Average rivers (ppb)	Small rivers	Average or larger rivers	Maxi-mum	Mid-range	Mini-mum	Small rivers	Average or larger rivers
Coke ovens									
No dephenolization	11.6	84.8	1.4	14	14	<<1	0	n/e	n/e[c]
With dephenolization	0.24	1.8	0.9	9	<1	<<1	0	n/e	n/e
Phenol manufacturing	12.6	94	0.5	5	5	<<1	0	n/e	n/e
Petrochemical	45.3	383	10	100	200	2	0	[b]	n/e
Oil refineries	4.4	35	10	100	20	<1	0	n/e	n/e
Plywoods	0.001	0.011	10	100	<<1	<<1	0	n/e	n/e
Blast furnaces	0.075	0.72	9	90	<<1	<<1	0	n/e	n/e
Phenolic resins	20.8	154	1.5	15	30	0.3	0	n/e	n/e

[a] These values are only for use in the analysis and should not be cited for other purposes.

[b] Above detection.

[c] n/e = No effect.

C. Cumulative Noncarcinogenic Chronic Risk

Phenol health effects data, as discussed, are incomplete and somewhat contradictory. For example, people have survived levels higher than those reported to be lethal. (This may be a function of compensation by the body for gradual exposure). Nevertheless, effects were not seen until concentrations reached parts per thousand. Estimates from the exposure and the river models, even for large populations at risk, show no human risk from exposure to only phenol from industrial sources. Furthermore, for average rivers, we should not expect effects at the levels shown in Table 34.

We estimate exposures from the use of medicines such as mouthwash, lozenges, lotions, and liquids applied to the skin to be about 0.4 g/day for adults and 0.03/day for children (see Appendix III-A). These values have negligible effect on risk estimates. On this basis, there are no chronic effects for nonmedicinal users since medicinal users exceed our worst case assumption for the former by five to ten times. That worst case assumes a petrochemical plant on a small river with effluent of 45.3 ppm phenol is added to 2.8 ppb/day from food sources, as shown in Table 34.

In summary, phenol and cadmium risk estimates confirm the initial reasoning for their selection as test substances for our analysis. All evidence considered supports the conclusion that phenol in water presents a very low health risk and cadmium in water may be a moderate health risk.

Appendix V-A

TRENDS IN CADMIUM USE

The transportation sector of the U.S. electroplating industry demands 17 to 25% of the total available commercial cadmium. The entire electroplating industry consumes 50 to 55%. Because of the reliability, long life, and low maintenance requirements of the Ni-Cd battery, this industry is rapidly increasing its cadmium demand.

On the other hand, demand for cadmium pigments (primarily used in plastics) should remain stable unless there is a shortage of oil and plastics or cadmium environmental problems cause a decrease in use. This is also true of cadmium heat stabilizer use in plastics. Other current use areas which may expand include both energy (solar photovoltaics, nuclear power generation, railway/trolley electrification) and electronics (TV surveillance devices, radar, visual display terminals). Current major consumers are listed in Table 2-A. Note that consumption in respect to the plastics and the pigments industries is interrelated.

Current use is expected to rise. Major U.S. consumers may require a yearly average of 6700 ton of cadmium by the year 2000 (Table 3-A).[151] This projection was developed by relating economic indicators such as GNP, gross private domestic investment, population, and industrial production indexes to historical data by regression analysis and then considering numerous contingency factors. Factors which may further increase demand are the expansion of mass transit systems, resolution of environmental problems, and increase in the number of users of extent of use. The lower end of each range assumed:

1. An increase in substitues due either to performance or environmental considerations of the cadmium and alternative product
2. Petroleum and energy shortage

Of the inadvertant emitting industries, the phosphate industries (primarily the phosphate fertilizer industry) have the worst total emission.[66] This may become even more significant as the use of phosphate fertilizers is expected to increase in step with crop export demands and continuing and expanding nutrient replenishment needs of agricultural land.

Demand for phosphate detergents, on the other hand, is decreasing as acceptable substitutes become available. It has been estimated that by 1985 over 50% of the U.S. population will be using substitutes for phosphate detergents due to intensive Federal and state regulation.[149] We have, however, apparently entered a period of deregulation and such prognostications are not now as easily made.

There are environmental concerns for cadmium pollution in the secondary metals industry for both nonferrous and ferrous metals. With regard to nonferrous metals, the secondary zinc and copper industries are a potential emitter of cadmium as an impurity of zinc and copper alloys. Secondary ferrous metals, i.e., scrap, enter the primary iron and steel industry in the form of cadmium electroplating or contaminated galvanized coatings.[65] Of special concern is the use of No. 2 scrap since approximately 0.045 kg/kkg, is completely volatilized in processing and may enter the environment via inefficiencies in dust-catching equipment, misuse, and/or inefficient disposal of the dusts.[65,66] The amount of cadmium emitted by the iron and steel industry depends on the amount of No. 2 scrap used and the efficiency of dust-catchers and disposal methods.

Table 2-A
PRIMARY
CONSUMPTION OF
CADMIUM[66,151]

Electroplating	50—55%
Plastics	12—20%
Pigments	12—13%
Ni-Cd Batteries	5—22%
Miscellaneous	2—8%

Table 3-A
CADMIUM DEMAND BY MAJOR
CONSUMERS IN THE YEAR 2000
(kkg)

Consumer	Range	Mean
Electroplating	700—1900	1100
Transportation	(700—1100)	(800)
Other	(100—1900)	(1400)
Batteries	1500—4500	2700
Pigments	800—1600	900
Plastics	500—1400	700
Other	150—380	200

Source: Bureau of Mines (Revised by AURA)

Appendix V-B

VALUE OF TECHNOLOGY - ADDITIONAL INFORMATION

I. CADMIUM

A. Mining and Smelting of Zinc and Cadmium

The closing of eight major U.S. zinc plants in the early 1970s and the continuing replacement of retort process plants with the newer electrolytic process plants has reduced cadmium environmental release.[65] The 1979 closing of the St. Joe Zinc Company plant in Monaca, Penn. alone may have had a considerable impact on U.S. cadmium production capacity as evidenced by the 32% decrease in production during the first quarter of 1980 compared with the previous quarter. This 1980 quarter production was also down 26% from the 1979 quarter total.[163]

This electrolytic method is considered a "cleaner" reduction process. In this process, the waste concentrate and collided flue dust are taken into solution with sulfuric acid, and a zinc-cadmium sludge is precipitated by adding zinc dust. Final recovery involves dissolution of the cadmium-bearing feed material, followed by various purification steps. The resultant solution contains less than 0.2 mg/ℓ cadmium. Because the spent electrolyte containing no cadmium is recycled as leachate, significant water-borne wastes are unlikely unless, of course, there are accidental releases. However, not only is the electrolyte recovery process cleaner, it is also less labor intensive and hence more efficient. This process could allow U.S. producers to maintain a 40% share of U.S. demand.[151]

Currently, approximately 72% of the cadmium is imported in one form or the other.[151] While overall U.S. cadmium demand is increasing 2.2%/year, U.S. production is decreasing 10 to 20%. Some attribute a substantial part of the decline to what have been increasingly stringent pollution standards. Regardless of cause, the immediate effects appear to be an increase in importation with some indication of a reduction in primary user, particularly in the electroplating industry.

Cadmium is included in the U.S. mineral stockpiling program. In 1976, the stockpile goal was set at 11,204 tons with the 1976 level at 2871 tons. There have been no subsequent releases or acquisitions of cadmium since 1976.[151]

Refined cadmium is marketed in several forms. Electroplaters prefer cast ball shapes; chemical and pigment users, sticks or flakes; alloyers, slabs and ingots. Producer prices range from $3.00 to $3.25/lb, considerably more expensive than zinc at $2.70/lb.[164]

B. Ni-Cd Batteries

There are two distinct types of nickel-cadmium cells: the pocket plate cell and the sintered plate cell. The majority of U.S. production is of the latter type. Sinter-plate Ni-Cd batteries contain a cadmium anode, a potassium hydroxide electrolyte, and a nickel oxide cathode. Sintered placques containing the active materials are used for the electrodes. Three processes may be used in preparing these electrodes: impregnation-sintered plate, electrolytic deposition, and pressed powder. The first two utilize water washes, the latter is a dry process.

The impregnation-sintered plate process is the most widely used. In this method, after hydroxide precipitation, the majority of the cadmium-nitrate settles as sludge and less than 3% of the water effluent concentration is discharged to sewers.[65]

Increasingly, nickel-cadmium batteries are being imported. In fact, a major part of U.S. Ni-Cd battery consumption depends upon imports. In fact, only ten plants in the U.S. now produce these batteries. (The four largest are located in Florida and Texas.) One reason is that the initial cost of manufacture is two to three times that of lead-acid batteries.

Sealed nickel-cadmium batteries vary from small button to cylindrical in shape and are used in calculators, radios, pacemakers, telephones, tools, and appliances. Vented nickel-cadmium batteries are larger and are used in aircraft, submarine standby power, diesel buses, and emergency lighting. Feasible future uses of the Ni-Cd battery include power plant load-leveling and wind and solar electric-power storage.[151] If, in the future, other costs continue to rise and conservation and environmental concerns become valued more highly, we might expect the Ni-Cd battery to achieve widespread acceptance.

C. Pigments

There are disadvantages to substitutes for cadmium pigments. For example, the available inorganic pigment substitutes for chrome yellows and oranges have about the same color range as do cadmium pigments, but their high lead content makes them even more toxic. The ferrite yellows and oranges are competitive in terms of cost but not color intensity. Titanium-dioxide tends to be more pastel yellow than is desirable.

There are also problems encountered in using organic substitutes. Organic yellows and oranges are relatively expensive when compared with inorganics. They tend to bleed and the majority do not have sufficient light fastness and durability.

The cadmium lithopone reds and maroons could be replaced by iron oxide and organic pigments. However, while iron oxides are inexpensive, they lack the color brilliance of cadmium lithopones. The organic reds and maroons are expensive and weather poorly. Briefly, the advantages of cadmium pigments are

- Heat stability up to 600°C
- Resistance to H_2S and alkalis
- Light stability when used as a solid color as well as when mixed with white
- Good hiding power and color intensity
- Permanence (it does not bleed)[165]

D. Electroplating

Cadmium has some drawbacks to use in electroplating, being significantly more expensive than zinc, but no better at temperatures above 500°F. In addition, it embrittles titanium and is toxic. An ideal substitute would, of course, be one which was plentiful, inexpensive, easily applicable by current technology, nontoxic, and have the desirable properties exhibited by cadmium. Paints are possible substitutes for the electroplating process when used on decorative objects and whenever temperatures are less than 500°F, or where low stress and wear are likely. Phosphate coatings and oil are used on many of the common ''nuts and bolts'' type hardware.

The major current substitutes are other metals. Of these, iron is unacceptable. It is, in fact, one of the major metals protected by cadmium in electroplating. Silver, gold, and the platinum metals might be satisfactory in terms of effectiveness but are simply too expensive for most uses. Of all the substitutes, aluminum is by far the most abundant.[60]

Another possible problem in substitution is the fact that zinc, in large quantities, and some nickel and chromium compounds, are toxic to man. In addition, useful properties of the substitutes do not completely span the range of those of cadmium. For example, only zinc and lead are acceptable substitutes to recover and separate the coating from steel. In respect to solderability, only copper is acceptable.

While aluminum can be substituted for uses such as the manufacture of functional and decorative hardware and guns, and zinc may be used in some electronic apparatus, cadmium is, nevertheless, superior to both in solderability, electrical conductivity, and electrochemical protection. In the transportation sector, the fastest growing sector of the electroplating industry, it is required because it confers quality, fuel efficiency, and safety. Cadmium, therefore, cannot be completely replaced by either zinc or aluminum.

E. Plastic Stabilizers

Lead stabilizers would be a satisfactory substitute for cadmium in plastic were it not for toxicity and the fact that not all of these stabilizers can be used in translucent PVC compounds. Metal salts and soaps, with the exception of tin, are predominantly either heat *or* light stabilizers. Some of these compounds are desirable in that they are nontoxic. On the other hand, some are stained by sulfides in the air.

Tin salts and soaps, primarily those containing sulfur, are the most likely substitutes, but they are more costly and are neither as light nor as weather durable as barium-cadmium. They are not appropriate for use in flexible PVC compounds, organic stabilizers, epoxidized oils, and benzophene light absorbers because of higher cost and tendency for yellowness. The major detriments to substitution are either toxicity or poorer performance followed in importance by higher cost.

F. Phenolic Resins

The primary resin-using industries are plywood and fiberglass insulation. For the former, there is no direct substitute for phenolic resin because of its unique qualities and lower costs. In fact, 98% of adhesives used are phenolic resins. However, the Forest Service has suggested that 30 to 50% of that used could be replaced by lignin, a natural glue.[98] In this case, lignin is not a direct substitute, but is combined with phenolic resin as an extender. Most phenol wastes are produced when glue-mixing equipment is washed. The effect of this substitution would be to reduce phenolic resins consumption by about 11% and that of phenol by about 5%.

Approximately 83% of the total insulation industry is engaged in making fiberglass insulation. Using nonglass fiber insulation would eliminate the need for phenolic resins as bonding agents. This is the only way to approach this particular problem since there is no direct bonding substitute. If this were done, phenolic resin consumption would drop about 20% and that of phenols about 9%.

We have no information on substitutes for other phenolic resin uses. Since other resins can be used for modeling compounds, it is conceivable that most phenolic resins for this use could be replaced. If so, phenol use would be reduced by about 3%. Overall, we estimate that phenolic resin use could be diminished by 35% and that of phenol, 16%.

1. Bisphenol A

Approximately half the bisphenol A is consumed in the manufacture of epoxy resins. Most of the remainder is processed into polycarbonate resins.

Allyl chloride is another major feedstock of epoxy resins. The proportion of bisphenol A currently made from allyl chloride is unknown, but many epoxy resins could conceivably be produced from that compound. If epoxy resin use were reduced 50%, phenol consumption would drop about 5%.

Polycarbonate resins are used in a variety of applications primarily because of their attractive chemical and physical properties. They also meet environmental standards more readily than do other resins. In fact, for many uses epoxy resins may be replaced by polycarbonate resins.

2. Nylon Fiber

Approximately 48% of nylon 66 is produced using caprolactum as a feedstock and the remaining 52% uses cyclohexane. If cyclohexane totally replaced caprolactum, phenol use would be reduced 14%. Demand growth for caprolactum has, in fact, slowed to 7% from 12% 5 years ago.

There are other subsitutes. Adipic acid, another phenolic compound, can also be used to make nylon 66. Currently, 98% of this synthetic is produced from cyclohexanol. Substitutes of adipic acid would reduce phenol usage another 28%.

Nylon is used primarily in the production of 75% of all carpet fiber. Most of the remaining 25% is produced for natural products such as wool. Nylon cannot be economically replaced in the main because of the high cost of natural fibers. The same problem exists in the knitted fabric industry, where cotton and wool are more expensive than synthetics, some of which can be made from other synthetics. Examples are the acetates used in the manufacture of warp knits.

Nylon plastics, on the other hand, comprise only 1% of all plastics.

Part VI

CLASSES AND TYPES OF STANDARDS FOR CONTROLLING WATER EFFLUENTS

Delay is preferable to error.

Thomas Jefferson
Letter to George Washington (May 16, 1792)

*Ignorance is preferable to error; and he is less remote
from the truth who believes nothing, than he who believes what is wrong.*

Thomas Jefferson
Notes on the State of Virginia (1781-1785) Query 6

There are things worse than error, but delay and ignorance are seldom among them. In order not to repent at leisure, one should realize that time taken to construct or adapt to our use an appropriate and practical model is time well spent.

In this part, we discuss and define inputs to an evaluation model, such as how the standard might be implemented, the desired level, alternate controls, and margins of safety. This model is used later to evaluate alternate methods of setting levels in standards for each source of water-borne risk from cadmium and phenol. We also describe the classes and types of standards.

Chapter 21

CONTROL STRATEGIES AND TYPES OF STANDARDS

If a man begin with certainties, he shall end in doubts;
but if he will be content to begin with doubts,
he shall end in certainties.

Francis Bacon
The Advancement of Learning (1605)

Our objective is to analyze and evaluate a range of methods for establishing risk levels for environmental standards. Before doing so, we must select a strategy to determine the kind of standard that will best serve under each control condition.

Assuming the adverse health effects of our "strawmen" pollutants in water occur with reasonable margin of error, the problem is how, and to what level, to control these industries. There are two basic strategies which may be followed, prevention or mitigation.

I. DEFINITIONS — STRATEGIES

Preventive strategies are, by definition, used before exposure occurs. They include controlling the source by effluent standards, restricting use levels by water quality standards and treating finished drinking water. They involve both monitoring and enforcement. *Mitigation strategies* are used after exposure or possible exposure. For example, one may deal with acute exposure from accidental discharges after an accident by breaking the pathway to man. This generally entails temporarily restricting consumption of water from a particular source. Health impacts, however, can result from cumulative intake from all sources, requiring certain specific mitigation strategies depending upon the nature of the toxin. For cadmium and other heavy metal poisoning, there are therapeutic measures available using chelating materials. Treatment with ascorbic acid or zinc diet additives can reduce cadmium metabolic effects. The treatments, however, especially use of chelating agents, may be quite uncomfortable and dehabilitating and are not without risk themselves.

Therefore, preventive strategies are usually preferred, the most obvious being control at the source through standards, surveillance, and enforcement. Other preventive approaches include restricting use of contaminated water by water quality standards, controlling concentration of finished drinking water by either removing the contaminant or diluting with clean water, and using prophylactic techniques.

Appendix VI-A contains definitions of the various generic forms of standards, i.e., performance, design, acceptance, and achievement, and discusses how they may be related through the use of models. Regardless of the generic form of a standard, an environmental standard may fall into one of several different classes based upon the particular opportunities for control.

II. CLASSES OF ENVIRONMENTAL STANDARDS

Classes of environmental standards may be divided into three basic categories as shown in Table 35. Each category may be further divided by where control is directed.

There are a variety of types of standards under each class heading as shown in Table 36. This list is not exhaustive; we have presented it only to illustrate that most of these methods are achievement standards (see Appendix IV-A for definitions). The two exceptions occur when there is a need for risk performance levels when using a "bubble concept" across

Table 35
CLASSES OF STANDARDS
FOR WATER

Effluent standards
    Point sources
    Area sources
Ambient standards (water quality)
    Use levels
    Raw water
Finished drinking water standards
    Surface water
    Ground water

Table 36
TYPES OF STANDARDS FOR WATER BY CLASSES

I.   Effluent standards
      Point and area sources
      A.   Concentration at discharge point (performance-standard)
          1.   As the only parameter
          2.   As a function of water flow for dilution
          3.   As a function of downstream concentration, based upon measurement and/or models (diffusion, flow, settling, degradation)
          4.   As a function of time and use pattern (for area sources there may be seasonal flooding and production cycles)
      B.   Amount per unit time — for point sources (performance standard)
          Amount per unit time/area at use level (performance standard)
          1.   As the only parameter
          2.   As a function of flow dilution
          3.   As a function of downstream measurement
          4.   As a function of time and use pattern (continuous or intermittant)
      C.   Specification of control methods by use (performance and design standards)
      D.   Specification and/or restriction of production methods (design standard)
II.   Ambient standards (water quality) for both use levels and raw water
      A.   Concentration (performance-standard)
          1.   Use limits
          2.   As a function of time and change of flow, etc.
      B.   Bubble concept
          1.   Across contaminants (performance-standard)
          2.   Intra-contaminant (performance-standard)
III.   Finished drinking water standards for both surface and ground waters
      A.   Concentration (performance-standard)
          1.   Dilution/mixing
          2.   Limits
          3.   Synergistic effects (chlorine/phenols)
      B.   Concentration vs. exposed population
          1.   Number of users
          2.   Type of user

contaminants (to provide a risk "equivalency") and when setting drinking water standards which vary based upon number or type of user (e.g., large vs. small population, ground vs. surface water).

III. ADDITIONAL FACTORS TO CONSIDER FOR ENVIRONMENTAL STANDARDS

Each type of environmental standard has a different implementation problem depending

Table 37
IMPLEMENTATION ASPECTS OF STANDARDS

1.   Enforceability
   - A.   Measurability
     1.   Ability to monitor
     2.   Cost of monitoring
     3.   Train of evidence of violations
     4.   Ability to validate performance
   - B.   Legality
     1.   Legislative authority
     2.   Executive authority
     3.   Judicial history
   - C.   Degree of Motivation for Voluntary Compliance
     1.   Mix of voluntary vs. involuntary compliance
     2.   Motivating factors
2.   Equity
   - A.   Equal treatment under the law
   - B.   Special problem areas
   - C.   Uniform vs. graduated standards
   - D.   Fairness
3.   Administrative Costs of Implementation
   - A.   Enforcement
   - B.   Training and education
   - C.   Information dissemination
   - D.   Legal defense

upon the specific approach. These are discussed in detail in Appendix IV-B, Implementation of Different Types of Environmental Standards.

In general, factors affecting implementation of environmental standards include the degree to which a standard can be enforced, as well as its legality, costs, and equity. Table 37 lists a number of these factors which must be considered, irrespective of the form of a particular standard. We have not attempted to deal with these additional factors in detail in our analyses.

If one or more of these aspects creates major problems, even standards which provide good performance or design measures may not be easily carried out. The choice of type of standard depends upon these factors as well as direct and indirect costs of the industry and the private and public sectors.

Chapter 22

THE EVALUATION MODEL

All human error is impatience, a premature
renunciation of method, a delusive pinning down of a delusion.

Franz Kafka (1884—1924)
Letters. Quoted in Max Brod.

Having considered strategies as well as the classes and types of standards available, we may now turn our attention to a model to evaluate various methods of setting risk levels for each standard. These methods are explored in depth in subsequent chapters using the model outlined here.

This model is key to understanding the remaining analysis in this book. Note that it is specific to this particular analysis.

I. STRUCTURE OF THE MODEL

The function of this model is the oft repeated purpose of this analysis: to evaluate methods for setting risk levels in standards for the various risks posed by cadmium and phenol in water.

Figure 25 graphically depicts the elements of the model. The box with the double outline band labeled "level of standard" is the heart of the model. It is here that we both initially determine the levels to set and later, by feedback, determine if those initial levels were best suited to the purpose,

There are actually three phases in the determination:

1. Propose levels of a standard by alternate methods for establishing standards
2. Determine how those levels impact a number of evaluation criteria
3. Evaluate the proposed levels by these criteria

A. Phase One

Phase One is begun by referring to a box (at the top of Figure 25) labeled "Sources of Risk". This is the first determination. Once the risks to be addressed have been identified, two paths are followed. In one, appropriate effects/impact models are used to give the risk and the value of technology which will be foregone if these risks are to be controlled. These models relate change in levels of exposure (impact of a control strategy) to changes in risk (effect). This information is obtained for each industry and each source of risk by models in Chapter 11, and is one of four inputs to the initial determination of level of standard.

The second path is the one we have embarked upon in Part VI by identifying (in Chapter 21) the class (Table 35) and type (Table 36) of available environmental standards for water. (A more detailed discussion of the latter appears in Appendix VI-B.)

Thus, for each of these sources of risk from cadmium and phenol one must look at all three classes of standards: effluent, ambient, and drinking water standards. As shown in Table 35, the classes of use can be further categorized as to where the standards are directed. Effluent standards, for example, may be focused on either point or area sources. Here, except for the phosphate fertilizer use, we will evaluate noncarcinogenic chronic cadmium risk from point sources only. Either point sources or ambient levels are used to determine cancer risks from cadmium and phenol, depending upon where control is exercised.

For effluent standards three different types of standards are evaluated, control of concen-

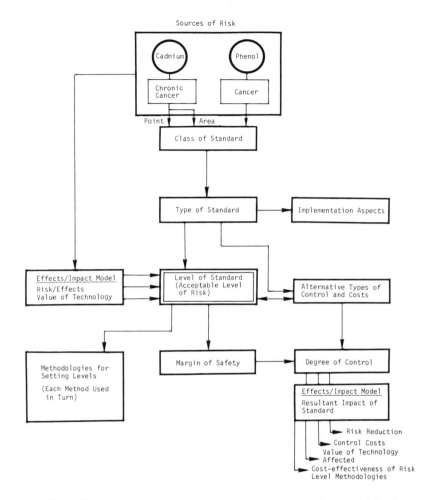

FIGURE 25.   An evaluation model for evaluation of methods of setting risk levels.

tration at the discharge point, as a function of river flow and measured downstream concentration.

As indicated in Figure 25, the implementation aspects of each type of standard (described in Chapter 21 and listed in Table 37) should be considered. These factors, which possibly hinder standard implementation, impact the level of risk selected for a particular type of standard. Of the types of standards available, only three types of effluent standards were chosen for further evaluation, since other approaches were not as easily implemented when the aspects tested in Table 37 were applied. The types of control available for standards and cost for a particular type of standard is yet another input. Table 38 presents a range of possible control options for effluent control.

There remains a fourth and final input before levels of risk for a standard can be proposed (completing Phase One of the model); that is, the methods for establishing acceptable levels of standards as shown in Table 39. Each method shown in Table 39 is applied in turn and the results evaluated,

A particular control results in altered exposure of the public. With the controls selected and the information from the effects/impact models, we can now determine both the effectiveness of a control in reducing risk and whether that control could stop or alter production (value of technology foregone).

Table 38
OPTIONS TO CONTROL EFFLUENT

A. Shut down plants — zero discharge
B. Shut down under adverse conditions — temporary
    1. Flow for dilution inadequate
    2. Accumulation of volume for controlled concentration of discharge
C. Waste material temporary storage
D. Apply control technology
    1. Retrofit
    2. New
E. Industry geographic concentration limits — co-location problems
F. Restrict location
    1. By river flow
    2. Juxtaposition to users
G. Reduce use (need) for product
    1. Substitutes
    2. Elasticity of demand (tax, surcharge, etc.)
H. Prophylaxis.

Table 39
METHODS OF ESTABLISHING LEVELS OF STANDARDS
(ACCEPTABLE RISK LEVELS)

A. Risk aversion
    1. "Zero risk" — local; global
    2. As low as can be measured
    3. As low as can be controlled
    4. As low as background
B. Acceptable risk
    1. Risk comparisons
    2. Arbitrary risk numbers
    3. A set value for $ to be sought
    4. A set value for risk reduction
C. Economic approaches
    1. Marginal cost of risk reduction
    2. Tax incentives
    3. Risk/cost/benefit balancing
D. Design approaches
    1. Best practical technology (BPT)
    2. Best available technology (BAT)

In summary, in Phase One, the effectiveness and expense of various control alternatives, as well as health impact and the value of technology foregone from the effects/impact model, are obtained for each of the methods in Table 39 to set the level for each of the classes and types of standards.

B. Phase Two

In this phase, we estimate the impact of the particular proposed level. But first, a margin of safety is set. This is a cushion obtained by lowering the level estimated to be safe to better protect all of the public. A margin of safety, usually required if a standard is to be acceptable, generally is a function of severity of consequence. For noncarcinogenic chronic risks, the consequences are some degree of illness, which may or may not be reversible. For these and for cancer risks, we examine two margins of safety, a factor of 2 and of 10, merely to illustrate their use.

Once one sets the degree of acceptability by using a margin of safety, this information and type of control are used to determine the degree of control that can be obtained for the required risk reduction.

Now we can evaluate, via the effects/impact model, how this degree of control impacts a number of criteria including risk reduction, control costs, value of technology foregone, and the cost-effectiveness of the approach.

C. Phase Three

Having evaluated all the methods of setting standards for each industry and risk source, the variation from the baseline and the sensitivity of the various criteria to uncertainty are used to determine which, if any, of the methods may be useful in setting acceptable levels for standards.

II. APPLICATION AND IMPACT OF CONTROLS

Ultimately, the appropriate effective degree of control for a specific case must be established. To do this one must examine how controls can be applied and the impact of each application. The degree of control selected was used as an input to the effects/impact model, which, in turn, provides a measure of impact.

For each application of a standard and method of setting a level of risk, residual risk level, degree of risk reduction, control cost, and value of technology foregone (all measures of the impact of control) are obtained from this process. These are shown in the lower right hand box of Figure 25. Cost-effectiveness of risk reduction on both incremental and marginal terms is an additional measure of impact computed from the above measures.

III. COST-EFFECTIVENESS CRITERIA

Both incremental and marginal cost-effectiveness measures serve as criteria to evaluate alternatives. When controls are applied sequentially, incremental cost-effectiveness is simply determined by dividing the number of health effects reduced by total costs, including those for both control and value of technology foregone. This provides per health effect reduced. Marginal cost-effectiveness is found by dividing change in cost between two controls (sequentially utilized) by the increment in health effects reduced by the additional control. Any application of control whose slope is positive rather than negative is eliminated as being inefficient. The difference beween these two methods is illustrated graphically in Figure 26. In the incremental case, we determine the total risk and costs independently for each set of controls. In the marginal case, we use only the change from the preceding control application.

IV. VARIABILITY IN MODEL PARAMETERS

Since there is considerable uncertainty in both costs and health effects, it is important to give attention to the ranges of variation. We assume that cost may be understated by a factor of 2; health impact overstated by a factor of 10 and understated by a factor of 2. On this basis, cost-effectiveness ranges from the base level to a factor 20 times higher.

Control costs could be annualized by using either crude annual costs or present worth. In this comparative analysis, we have used only the crude annual costs, based upon 1980 dollars. We used standard inflation factors in appropriately adjusting 1971 and 1975 costs. Based upon the producers price index, the 1971 to 1975 correction factor is 2.15, for 1975 to 1980, 1.40.[166] Trial analysis has shown that present worth costs do not alter the conclusions of this study. Errors in cost are much smaller than these in either risk estimates or estimates of value of technology.

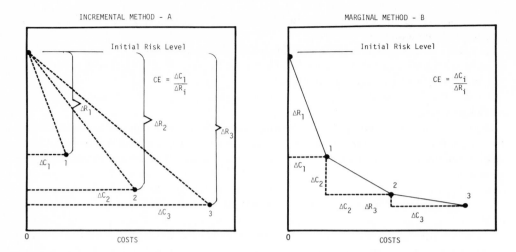

FIGURE 26. Cost-effectiveness of controls for health effect reduction. Controls 1, 2, and 3 are applied sequentially in each case resulting in same risks reduced and the same costs, but different cost-effectiveness values.

Appendix VI-A

GENERIC FORMS OF STANDARDS

Because there may be confusion over the meaning of some basic terminology with regard to standards, clarification may be necessary.

There are different forms of standards used in environmental protection and control systems. One form, the acceptance standard, specifies an acceptable level of risk in service, either explicitly or tacitly. This requires a societal decision as to the needed protection and, in some instances, a weighing of the expense involved. Sometimes, it may be impossible to directly measure how well such a standard achieves its goals in actual practice.

A second form, the achievement standard, describes the achievable level of control (under testable conditions) which will meet protection requirements. With this form of standard, different requirement levels can be stipulated — for example, a particular effluent removal system that performs at specified levels of removal.*

Acceptance standards are set at a level where socio-economic factors are balanced against risks and must be set by individuals at risk or by public officials responsible for public safety. Again, as described above, they may be set in terms of performance (level of risk to be achieved) either directly or through surrogate measures, or in terms of design, which specifies configurations to meet established criteria. Models relating risk to the measure employed may be used as surrogates. We will call these effects/impact models. When a performance level cannot be measured directly, the models provide a means of test methods to represent actual conditions.

Independent of whether a standard is of the acceptance or achievement type, a *performance standard* specifies the performance a system must achieve, i.e., a specific level of risk or a not to be exceeded mean time to failure greater than 100 hr. Similarly, a *design standard* specifies the technical design to be used; for example, that which will give a d.f. factor of 100 or a specific control method. A design standard codifies properties such as dimensions, rating, accuracy, performance capabilities, and operation under varying conditions. Thus, as we have said, both acceptance and achievement standards may be either performance or design standards. This is illustrated in Table 4-A.

If a standard calls for a not-to-be-exceeded level of either risk or exposure to recipients, it is termed a performance standard of the acceptance type. A standard specifying a given concentration at an outflow is also a derived (or surrogate) performance standard, related to risk by an effects/impact model. These models are shown in the left-hand column of Table 4-A. The river exposure models are examples.

The required use of "Best Available Technology" or "Best Practical Technology", as specified in the Clean Water Act, shows how design standards are used at the acceptance level. With this type of standard, one can determine the performance level achieved by a particular design. For this, translation models may be used. This is indicated in the third column of Table 4-A. Note that these models also work in reverse. They relate technology to the resulting required performance and risk levels. Either a particular type of control or a specified decontamination factor (both design standards) could become a performance standard via a translational model. Such models require a degree of standardization.

There are hierarchies of design and performance standards. For example, specifying a particular control method can achieve a given decontamination factor. That is, the d.f. is designed to meet a performance level (concentration) which now becomes a performance criterion to be met by specific control methods.

---

* A specified level of removal is sometimes referred to as the decontamination factor (d.f.).

Table 4-A

TYPES OF STANDARDS AND MODELS FOR RISK CONSIDERATIONS

Type of standard	Models/Specification			
	Model	**Specification**	**Model**	**Specification**
	Effects/impact models ◄──►	Performance standards (operational specification) ◄──►	Translational models ◄──►	Design standards (technical specification)
Acceptance (Socio-economic judgment)	(Socio-risk-economic) models Dose-effects-risks Exposure Populations Control effectiveness Control costs Value of technology direct; indirect Social impact Intangible impacts	Level of risk in terms of measurable surrogates — set by regulators	Technical models Translate between design and performance	Level of risk in terms of specific designs
(Regulator) Interface ◄──► (Regulatee)				
Achievement (Technical-economic judgment)	(Technical-economic) models Measurement capability Reliability Alternatives/ tradeoffs Cost/Performance Effectiveness	Level of performance to be achieved by regulated party	Technical models Translate between design and performance	Design to be achieved by regulated party

In spite of this complexity, the point to remember is that levels of risk are best set by acceptance standards of the performance type. These levels of performance cannot easily be directly measured. Thus, design standards of the acceptance type are often substituted and translational models are used to relate the two conditions.

Once an acceptance standard is established, it becomes an achievement standard to be met by the regulated. This is shown in the bottom row of Table 4-A. These standards do not involve social judgments but address only the needed technical capability to attain either the performance or design specified. If the standard is not achieved, enforcement can be either at performance or design levels since one may measure compliance directly in most instances. The interface between acceptance and achievement standards represents the border between social value judgments of both safety and environmental regulation on the one hand, and technical aspects of operationally meeting regulations on the other.

Appendix VI-B

IMPLEMENTATION OF DIFFERENT TYPES OF ENVIRONMENTAL STANDARDS

Here we discuss the various types of standards and their application to water-borne pollutants. The three major classes of environmental standards, as described in Chapter 21, are

1. Effluent standards
2. Ambient standards (water quality)
3. Finished drinking water standards

I. EFFLUENT STANDARDS

Point sources, and in some cases area sources, can be controlled by effluent standards. Consider concentration at the discharge point as an example of a means of establishing a standard. Here, the objective is to maintain an acceptable risk level to exposed populations downstream. This performance objective can be related to a derived concentration standard through the exposure model (effects/impact model) acting as a translational model. This standard would set a maximum limit on concentration at some point downstream. This can be measured directly, but if there is more than one effluent source discharging into the same body of water, the sources of contamination and the contribution of different sources cannot be easily ascertained for enforcement. Moreover, wasteload allocations which might be required among present and future users are not easily resolved. Therefore, the derived standards can be related further back to another performance standard set at the discharge point. The exposure model is required for this relationship in the form of a mixing and dilution model. A variety of derived performance standards are possible.

A. Concentration at the Discharge Point

A maximum allowable concentration at a discharge point provides an enforceable, equitable approach to controlling effluents on a legal basis. Since concentrations may vary widely over time, the application of a standard to meet performance requirements under all conditions may be ineffective and costly, especially because discharge into low-flow streams is the major problem for chronic risks.

1. Concentration at the Discharge Point as the Only Parameter

One can set a concentration level at the discharge point which is the same for all plants. A concentration set high enough to allow plants on low-flow rivers to operate could presumably allow other sources on larger streams to have higher discharges than necessary. The acceptable level of concentration at the pipe is the governing factor in terms of effectiveness of controls and costs. All plants would be regulated the same way and the standard could be enforced by directly measuring discharge concentration. Control of concentration does not, however, limit the total amount of contaminant discharged.

2. Concentration as a Function of Water Flow for Dilution

Another approach is a discharge pipe concentration standard which is variable up to a maximum level with the volume of water flow per unit of time for mixing and dilution. Water flow is the independent variable in a model for mixing and dilution. To meet such a standard, plants on low-flow rivers would require controls not necessarily needed by other plants. The objective is to meet the performance level downstream and requires measures of stream flow at a given time and concentration of discharge for enforcement. Those plants located on low water flow rivers might consider the standard as inequitable, requiring them to install controls where those with high dilution flow need not. However, in this case, those plants which require control expenditures are those causing the risk. They include how to handle:

1. Apportionment schemes for multiple dischargers
2. Control after the fact
3. Time delay between exercising controls and their effect

In addition, enforcement is difficult, since actual exposure is measured at the downstream point. The objective is to use a water quality ambient level as a means to apportion and control discharges and, as a result, also control risks at the source as needed.

Models which may be employed would consider diffusion, flow, settling, and degradation of the pollutant.

B. Variable Standards Based upon Functions of Time and Use Patterns

Using concentration or amount limits which vary as a function of time is a means of controlling discharge for widely varying conditions. Note that area sources, particularly, may be subject to factors causing wide fluctuation in concentration and amount of discharge. These include flooding and both seasonal and nonseasonal high and low production swings.

1. Amount per Unit Time at a Discharge Point or Area Sources

Amount per unit time method is a way of controlling the actual level of contaminant discharged for both point and area sources. When using this method for area sources, the size of the area one is considering also becomes an important variable. However, if discharges have very high concentrations, inadequate mixing and dilution may result. To prevent this, one may control both amount and maximum concentration. The problem is that a decision must be made as to whether to allow each plant the same amount of discharge independent of mixing water availability. Obviously, if the same amount gets into a low-flow stream as into a higher flow stream, there may be risk in the former but not the latter. A level set low enough to prevent any risk might severely limit production or involve unnecessary controls on an average or high-flow stream.

2. Concentration as a Function of Time

Concentration as a function of time is a means to control the amount of discharge over a given time period. It is not particularly applicable in this situation.

C. Amount or Volume as the Only Parameter

It is difficult to measure and enforce discharge of the specified amount for either point or area sources. The volume of material released is a more likely manner of estimating area source pollution. Amount may often by determined by using the discharge volume per unit time and concentration. Concentration is measured directly, amount indirectly. One may, however, consider a number of other variants as we did for concentration.

1. As a Function of Flow Dilution

This method has all the advantages and disadvantages as the use of concentration but, in addition, provides for an absolute limit on contaminant discharged. Measurement, however, may be indirect. Some plants on low-flow streams will be restricted on the amount of pollutant which may enter the water.

2. As a Function of Downstream Measurement

This is the same measure as for concentration but limits the amounts discharged.

D. Specification of Control Methods by Use

Control methods may be specified for both performance and design levels. A performance standard might stipulate the removal fraction or decontamination factor while a design standard identifies the particular control method, e.g., a double liming process. The degree and choice of design of a performance standard is based upon acceptable risk levels, cost balancing, and reliability of control performance.

There are at least two possible ways to control:

1.   Effluent stream cleanup
2.   Temporary holdup during low flow

One may use the latter here, since low flow causes this particular problem. Control installation and operation is easy to monitor and enforce. However, uniformly applying such controls may be ineffective in other than low-flow locations.

E. Specification and Restriction of Production Methods

Certain types of mining and smelting processes, captive electroplating methods, and phosphate mining and production methods may have less, or more easily treatable, effluents than do other processes. Requiring the use of processes with lower effluents which may be more efficiently controlled is an alternate means of control. Conversely, shutting down particularly poor processes is also possible. Observing the degree of compliance in these cases is sraightforward, yet enforcement may be difficult since replacing processes or re-stricting use of existing facilities may conflict with other legal restrictions.

F. Summary

As we will learn in upcoming chapters, the most obvious choices for setting effluent controls for chronic cadmium risks are

● Concentration and amount discharged
● Concentration and amount discharged based upon flow and mixing models

In addition, consideration must be given to low-flow sites as opposed to high-flow sites for:

● Differential specification of removal fractions
● Specific methods of treatment based upon flow

II. WATER QUALITY STANDARDS

Water quality standards relate measured ambient level concentrations to particular uses. For example, high cadmium residues in raw water might restrict untreated drinking water ingestion, irrigation (to prevent build-up in foodcrops), and shellfishing. Such restrictions can be permanent or temporary, as when there is a low flow during drought or other unusual circumstances.

Ambient cadmium levels may result from many sources, including a variety of industry dischargers, area runoffs, release of sediment-held cadmium by dredging or acidity change, and deposition through fallout from air pathways. This problem may be mitigated somewhat since it is possible to directly measure ambient levels and restrict end use. Control of new and existing sources to maintain the ambient level for a given use is a complex undertaking in many cases.

For cadmium and phenol, the problem is on low-flow streams. A water quality standard is generally not useful in these cases, but one temporarily restricting use can be effective in reducing risks from accidents, illegal releases, juxtaposition of many sources, and faulty models (both river flow mixing and dilution). Ambient levels of cadmium and phenol are generally very low, and need only be addressed where discharge is to low-flow rivers.

III. FINISHED DRINKING WATER STANDARDS

If heavily contaminated raw water is used as a drinking water source, residues in the finished water may be removed or diluted by mixing. If, in this situation, ambient levels are low, the burden is on the community rather than the industrial sources. Conversely, if ambient levels are high, the cost of obtaining finished drinking water depends upon control costs and the availability and expense of obtaining water from an alternate source. As we have observed, there is little cadmium or phenol in most drinking water, since finished drinking water standards are well below chronic risk levels.

Part VII

EVALUATING APPROACHES TO STANDARDS FOR CADMIUM NONCARCINOGENIC CHRONIC RISKS

Order and simplification are the first steps toward
the mastery of a subject — the actual enemy is the unknown.

Thomas Mann
The Magic Mountain (1924), ch. 1

At this point, in the interest of order and clarification, it may be well to briefly review several points.

Acute risks of cadmium appear to result only from accidents or disregard of standards for chronic risks. Once a standard for chronic risks is established at any reasonable level, acute risks are simultaneously controlled. Noncarcinogenic chronic risks with a nonlinear dose-effect relationship are treated differently than those for cancer, where we assumed a nonthreshold dose-effect relationship.

Bearing these facts in mind, in Chapters 23 through 25 we evaluate the methods of setting risk levels for noncarcinogenic chronic risks of cadmium for each of five types of standards. Methods of risk aversion, acceptable risk, economic and design approaches are each addressed in separate chapters. Chapter 26 summarizes the findings. Cadmium and phenol cancer risk are discussed in the same manner in Parts VIII and XI, respectively.

For cadmium, there are three industries which present a significant noncarcinogenic chronic risk to humans in worst case situations. These results and the ranges of uncertainty for them are useful in this analysis, but may not describe the actual situation in a valid manner. If we had not found risks from our derived estimation, the results would have been of no value for our purpose.

Note that we did not consider in detail several aspects of possible risk. For cadmium, there are two such instances:

1.   Residues captured and retained in river sediment
2.   Residues deposited in rivers from sources other than fertilizer

Cadmium generally remains in sediment, although dredging operations, floods, and changes in water acidity could affect movement. In addition to reducing the amount of toxic and hazardous pollutants in water, the Clean Water Act (Sec. 304(b) and 306) mandates EPA consideration of the impact of this reduction on other problems, i.e., air pollution, solid waste management (sludge), energy requirements, and consumptive use of water. A consideration of these effects was beyond the scope of this project. The latter may particularly be an issue in the more arid western states.

Many of the figures and tables cited in this Part, as well as Parts VIII and XI, appear in previous chapters. This is to be expected, as we are now using previously generated information to assess, via our evaluation model, methods of setting levels of risk for the various polluting industries.

Chapter 23

RISK AVERSION METHODS — CADMIUM NONCARCINOGENIC CHRONIC RISKS

*People in general have no notion of the sort
and amount of evidence often needed to prove
the simplest matter of fact.*

Collected Works of Peter Mere Latham
(1789—1875)

In this chapter, risk aversion methods for the noncarcinogenic chronic risk of cadmium are discussed. For each method, we apply the various types of standards to each industry to see the extent to which various levels of noncarcinogenic chronic risks are met.

I. "ZERO-RISK" (WATER PATHWAYS ONLY)

Since the available literature suggests a no observable effect level for chronic exposure to cadmium in water of 120 ppm (maximum value D, Figure 15), a no risk condition is believed to exist below this level. Therefore, if annual exposure from all pathways is kept below this level with an adequate margin of safety, one may expect that no chronic effects will be manifested. Of course, when considering water pathways alone, zero risk applies only to water concerns.

An eventual onset of kidney disease in man occurs at exposure levels three to ten times higher than the no observable effect level. This effect can be reversed or offset, so that a large safety margin, in addition to the three to ten already built in, may be unnecessary. Monitoring problems here are minimal and pathways can be interrupted if the controls prove ineffective. Therefore, we selected as arbitrary alternatives two margins of safety, a factor of 2 and of 10, to illustrate how margins of safety impact levels of acceptable risk.

A. Effluent Standards
Now we can apply each method of setting risk levels for effluent standards to each emitting industry.

1. Mining and Smelting
For mining and smelting industries on small rivers, the downstream concentration at the model dilution point is 1031 ppb and for an average river, 8.14 ppb (Table 25). The average concentration at the pipe is 15 ppm prior to dilution. Of the seven plants, two are assumed to be on small rivers, affecting only one tenth of the total population. This is consistent with Table 30 and assumes that there is less population where water is scarce.

a. Concentration at the Discharge Point
To reduce the effluent of the two plants on low-flow rivers to a no observable effect level, we will need effluent reduction or decontamination factors of about 20 and 100 to provide the margins of safety of 2 and 10, respectively. Concentration at the discharge point is then correspondingly 0.75 ppm and 0.15 ppm. Control is required at all plants. By limiting pH to 10.5 and allowing settling, we may obtain a reduction factor of 30. This demands an initial investment of $250,000 per plant and $76,000 for yearly operation (1971 costs from Table 25). An additional reduction factor of 3 may be added by a 35% reduction in waste

---

SAFETY MARGINS

Margins of safety must take into account *at least* the following:

- **Varying sensitivities in the population at risk**
- **Uncertainty in health effect estimates**
- **Uncertainty in measurements for controls**
- **Uncertainty in pathway models**
- **Severity of the consequence of the disease**

    Some may argue that a *mininum* margin of safety of 100 is necessary if only animal data is used:

- **10X to account for species variation**
- **10X to account for individual susceptibility**

---

water volume to allow better downstream mixing. This will cost $176,000* in capital expenditure and $75,000 for yearly operation (again, based upon 1971 evaluations). Table 40, entry A1, summarizes expenditures necessary to achieve a zero risk standard based on reducing effluent concentration equally at all seven plants with cost adjusted to 1980 levels. We assume low-flow cases reflect long-term situations. If this were not so, less costly holding ponds could be used.

A factor of 100 reduction cannot be reached by available methods. If it were required, an industry valued at $16 million/year and employing 4600 workers (Table 12) would likely close. Assuming a safety factor of 90 is adequate, the entire industry would incur capital costs of about $3 million and a yearly operating expense of $1 million. Note that all standards can achieve "zero" noncarcinogenic chronic risk from water pathways alone. This can prevent an estimated 125,000 cases of kidney disease annually at worst and 10% or 12,500 cases, at the limit.

Installing settling tanks and controlling pH results in a factor of 30 reduction and provides a margin factor of 3 over the required factor of 10 reduction. This means we would pay a range of $11 to $220 per each health effect reduced (for both incremental and marginal approaches). Achieving a reduction factor of 90 involves an incremental cost-effectiveness of $22 to $440 per health effect. This cost effectiveness is essentially infinite on a marginal basis because there is no further reduction in health impact. Plant closure would cost $130 to $2600 per health effect and eliminate 4500 jobs. On a marginal basis, control above that provided by the initial approach buys margins of safety or increased confidence, but results in no measurable change in health impacts.

b. Concentration as a Function of Flow

A standard of 15 ppm would be established at the pipe, based upon an adequate flow of dilution water *and* a model** (formula) which lowers concentration at lower flow so that no more than 1.5 ppm, the "no margin of safety" level, is allowed in low-flow situations. Only two of the seven plants are controlled. Results appear in entry A2 in Table 40.

Since flow of diluting water is controlling, further control of the amount discharged does not provide additional protection. Moreover, the models used for adjustment by flow could be extended if one decided to use downstream measurements to assure adequate mixing.

---

*   Cost and risk estimates are not precise to the three significant figures shown. Rather, they serve as identifiers for particular numbers and to help the reader in locating them on a table.

** The details of the model are not necessary here, but the model must be standardized.

Table 40

APPLICATION OF CONTROLS FOR CHRONIC CADMIUM HEALTH EFFECTS: ZERO RISK FOR EFFLUENT STANDARDS

Entry number	Industry and standard type	Safety margin	Reduction factor Required	Reduction factor Achieved	1980 Annualized control costs ($ million/year)	Value of technology	Health effects reduced Best estimate	Health effects reduced Range	Cost effectiveness Incremental ($/he)[a]	Cost effectiveness Marginal ($/he)[a]
A1	Mining and smelting									
	Concentration alone	—	10	30	1.3		1.25×10^4	1.25×10^4	11—220	11—220
		2	20	30	1.3				11—220	—
		9	90	90	2.6				22—440	—
		10	100	90		16 M	125 K	12.5 K	130—2,600	—
A2	Concentration	—	10	30	0.39		125 K		3.2—64.0	3.2—64.0
	vs. flow	2	20	30	0.39				3.2—64.0	—
		9	90	90	0.75				6.5—130.0	—
		10	100	90		5 M			40.0—800.0	—
B1	Captive electroplating shops	—	6	14	9.7×10^3		2×10^6	2×10^5	4900—97,000	4900—97,000
		2	12	14	9.7×10^3				4900—97,000	—
	Concentration alone	5	30	33	1.3×10^4				6500—130,000	—
		10	60	100	1.9×10^4				9500—190,000	—
B2	Concentration	—	6	14	9.7×10^2		2×10^6	2×10^5	490—9,700	490—9,700
	vs. flow	2	12	14	9.7×10^2				490—9,700	—
		5	30	33	1.3×10^3				650—13,000	—
		10	60	100	1.9×10^3				950—19,000	—
C1	Phosphate fertilizer	—	3	30	18.0		3.3×10^5	3.3×10^4	53—1,100	53—1,100
	Concentration alone	2	6	30						—
		10	30	30						
C2	Concentration	—	3	30	1.8		3.3×10^5	3.3×10^4	5.3—110	5.3—110
	vs. flow	2	6	30						—
		10	30	30						

Note: Mining and cost data used for phosphate plants.

[a] he = health effect

Since this is expensive on a long-term basis, it might be used only during the first year or two to verify model adequacy.

2. Captive Electroplating

Downstream concentration of captive electroplating shops on small rivers are 583 ppb; on average rivers, 4.5 ppb. Populations affected are 10 and 100 million, respectively (Table 30). Without control, average flow from the pipe is 15 mg/ℓ (ppm), as shown in Table 12. Of the 9450 electroplating shops, 6050 are captive and 10% of these, 605, are assumed to be on low-flow rivers. The latter are estimated to cause about 2 million instances of kidney disease annually (Table 30).

a. Concentration Level at the Discharge Point

To reduce effluents of the 605 plants on low-flow rivers to a no observable effect level would require:

- An effluent reduction factor of about 6 with no margin of safety
- A factor of 12 for a margin of safety of 2
- A factor of 60 for a margin of safety of 10

Table 24 shows that both chromium reduction and cyanide oxidation are needed prior to pH adjustment to attain a reduction factor of 14. Thus, a single shop would invest about $1.6 million/year on a crude annual cost basis. Clarification gives further reduction to a factor of 33 at an annual cost of $547,000 and reverse osmosis achieves an additional factor of 3 (100 total) for $954,000 annually.

Table 40, entry B1, indicates that discharge point concentration at all 6050 shops must be controlled. Costs and cost-effective values here are substantially higher than those for mining and smelting. Moreover, since any control costs exceed value of technology, control causes industry shut-down. About 130,000 employees would then be unemployed.

b. Concentration as a Function of Flow

As with mining and smelting, we will need both a maximum concentration standard in the order of 15 ppm for normal flow *and* a model (formula) requiring lower levels at diminished flow. Application of this approach shows that only 10% of the captive shops need controls. (See entry B2 of Table 40). Since controls are effective, value of technology is not changed. Nevertheless, industries on low-flow rivers cannot afford these expenditures and would selectively be forced out of business. About 13,000 employees would then be unemployed.

3. Phosphate Fertilizer Production

Downstream cadmium concentrations from phosphate fertilizer production plants on small streams are 306 ppb; on average rivers, 2.3 ppb. Populations affected are 3.3×10^6 and 33×10^6, respectively (Table 30). Without controls, the 92 plants have a mean 16 ppm concentration at the pipe (Table 12). We assume nine plants are on low-flow rivers. Effluent from these plants cause an estimated 330,000 annual cases of kidney disease.

a. Concentration Level at the Discharge Point

To reduce effluents of the nine plants on low-flow sites to a no observable effect level requires an effluent reduction of about 3 with no margin of safety. A factor of 2 margin of safety, requires a reduction of 6; for a factor of 10 margin of safety, a reduction of 30.

The same processes for controlling effluents in mining and smelting are appropriate for phosphate plants. Use of liming to control pH has a crude annual cost of $192,000 at 1980 levels, at a reduction factor of 30 (Table 25). Cost of waste water reduction is nearly the same but because of downstream mixing efficiency has a reduction factor of only 3. It seems that pH control is cost-effective on an engineering basis alone. The expense for all 92 plants was $17.6 million in 1980. This is shown in Table 40, entry C1.

b. Concentration as a Function of Flow

Adopting a standard of 16 ppm at the pipe as a maximum *and* a model based upon flow means that one must apply controls at only nine low-flow sites. This does not affect the value of technology. These entries appear in Table 40, entry C2.

B. Ambient Standards (Water Quality)

One alternative to control at the source is control of the ambient cadmium level in low-flow rivers, Here, use of water which exceeds ambient drinking water standards for agricultural use, and shellfish harvesting could be restricted without further treatment. Shellfish harvesting does not present a major problem on low-flow rivers, but could be in certain areas of the South where the aquacultural raising of crayfish is practiced and where there are many phosphate plants. The expenditure for either agricultural or aquacultural use is the cost of obtaining alternate supplies of water. This also applies to drinking water.

The measured ambient amount in water is about 0.4 ppb maximum, far under a zero risk level. In fact, to meet a standard based on zero risk, ambients could be 2500 times higher with no margin of safety (100 ppb), 1250 times higher with a safety factor of 2 (50 ppb), and 250 times higher with a safety factor of 10 (10 ppb).

The cost of restricting use is obtained by multiplying water needs of the total population affected in low-flow cases by a daily usage factor of 160 gal per person and cost per 10^3 gal from alternate supplies. The 160 gal figure includes a 55 gal/day household use per person (the remainder is from all other uses) in which a population of 60 million in the 100 largest cities used 9650 MGD.[167] The expense of replacement water is the cost of obtaining the water and transporting it a mean distance of 30 mi. Costs of supplying water are estimated at $0.50/$10^3$ gal for water rights and $0.90/$10^3$ gal for transportation; a total of $1.40/$10^3$ gal.[168] Note that expenses for clean water are not borne by the polluting industries but by users.

1. Mining and Smelting

The two mining and smelting plants on low-flow rivers affect a population of 250,000. An outlay of about 2×10^7 is needed for the annual substitution of 1.5×10^{10} gal of water. This would be necessary at any ambient level below 100 ppb since the downstream level resulting from effluent is 1030 ppb. Resultant values are shown in entry D1, Table 41. Control here costs more than effluent control; moreover, it would be paid by the general population. Expenditures are almost independent of the level at which the standard is set, down to the ambient level.

2. Captive Electroplating

Effluents from the 605 shops along low-flow rivers affect 10 million people. Again, the monetary outlay is to obtain substitute water for these people. If we consider the same ambient standards as for mining and smelting, 5.9×10^{11} gal/year must be replaced at a cost of 8.2×10^8/year. About 2×10^6 cases of kidney disease would be averted. Entry D2 in Table 41 summarizes the results. Ambient standards here, unlike those for mining and smelting, cost less and are more cost-effective than effluent standards. Although polluters would not normally be cost bearers, they could be taxed an amount equivalent to that which the general population would normally pay.

Table 41
APPLICATION OF CONTROLS FOR CHRONIC CADMIUM HEALTH EFFECTS: ZERO RISK FOR AMBIENT AND DRINKING WATER STANDARDS

Entry number	Industry	Safety margin	Standard	1980 Annualized control costs ($ million/year)	Value of technology	Health effects reduced — Best estimate	Health effects reduced — Range	Cost effectiveness — Incremental ($/he)[a]	Cost effectiveness — Marginal ($/he)[a]
Ambient									
D1	Mining and smelting	—	100 ppb	20.0		125×10^3	12.5×10^3	1.6×10^2 —	1.6×10^2 —
		2	50 ppb					3.2×10^3 —	3.2×10^3 —
			10 ppb						
		100	1 ppb						
D2	Captive electroplating shops	—	100 ppb	8.2×10^2		2×10^6	2×10^5	4×10^2 —	4×10^2 —
		2	50 ppb					8×10^3	8×10^3
		10	10 ppb						
		100	1 ppb						
D3	Phosphate fertilizer	—	100 ppb	2.7×10^2		3.3×10^5	3.5×10^4	8.2×10^2 —	8.2×10^2 —
								1.6×10^4	1.6×10^4
D4	All industries above	—	100 ppb	8.2×10^2		2.5×10^6	2.5×10^5	3.3×10^2 —	3.3×10^2 —
		2	50 ppb					6.6×10^3	6.6×10^3
		10	10 ppb						
		100	1 ppb						
Drinking Water									
E1	Mining and smelting	3	34 ppb	3.5		1.25×10^5	12.5×10^3	28 — 560	28 — 560
E2	Captive electroplating shops	5	20 ppb	1.4×10^3		2×10^6	2×10^5	700 — 14,000	700 — 14,000
E3	Phosphate fertilizer	10	10 ppb	46.0		3.3×10^5	3.3×10^4	140 — 2,800	140 — 2,800
E4	All industries above	3	34 ppb	1.4×10^3		2.5×10^6	2.5×10^5	560 — 11,200	560 — 11,200

[a] he = health effects

3. Phosphate Fertilizer Plants

Nine plants on low-flow rivers affect 3.3×10^6 people with yearly totals of 1.9×10^{11} gal of water, Controls would cost 2.7×10^8 and eliminate 330,000 cases of kidney disease annually. Results are shown in entry D3, Table 41. In this case, the expenditures are substantially higher than those for effluent control. Of course, our assumptions as to the price of alternate water sources and amount required are quite variable.

4. Total Ambient Control

It is possible, but not probable, that all industries with plants on low-flow rivers are on the same bodies of water. In that situation, the cost of restricting water use would be for that of the largest using industry. This would reduce health impact from all sources and is reflected in entry D4, Table 41. Both costs and cost-effectiveness are lower for other entries in the table. They would be even lower if the exposure pathways of other pollutants were simultaneously interrupted by such standards.

C. Drinking Water Standards

Neglecting for a moment the agricultural and aquacultural pathways, an alternate means of interrupting exposure pathways is to permit all water use but also clean up drinking water. This would provide safe levels of:

- 100 ppb (mg/ℓ) With no margin of safety
- 50 ppb With a margin of safety of 2
- 10 ppb With a margin of 10*

Cleaning up drinking water requires technology similar to that used for mining and smelting, namely, controlling pH and allowing settling. This provides a reduction factor of 30. We are using a 55 gal/day of drinking water rate per person. The process shown in Table 25 is based upon processing 370,000 gal/day. A plant of this size serves about 6700 people at a 1980 annual crude cost of $190,000, about $28.2 per person served. On a larger scale, we assume economies of scale could reduce average costs to $14 per person.

1. Mining and Smelting

Applying drinking water controls reduces the effluent level to about 34 ppb and provides a safety margin of 3. Results are summarized in entry E1, Table 41.

2. Captive Electroplating

Applying controls in this case results in a 20 ppb level with safety factor of 5. Results are summarized in entry E2, Table 41.

3. Phosphate Fertilizer Plants

Controls result in a 10 ppb level, with a safety factor of 5. Results are summarized in entry E2, Table 41.

4. All Sources Covered by Cleanup of Finished Drinking Water

We assume cleanup of water from the industry affecting the largest population (electroplating) would simultaneously reduce health impact from other sources. These results are shown in entry E4, Table 41. The standard, 34 ppb with a margin of safety of 3, allows discharge by the worst polluting industry (mining and smelting).

In this case, drinking water control costs less and is more cost-effective than control of

---

* The 1980 EPA water standard to protect human health is 10 ppb.

ambient levels by use. Of course, agricultural, aquacultural, and shellfish pathways are not controlled by ensuring clean drinking water.

If regulations insist on industry controls, it should be known that some would have to pay more. Mining and smelting and phosphate plants expenses would be considerably less than those of captive electroplating shops since smaller populations and fewer plants are involved.

D. Summary — Zero Risk (Water Pathways Only)

Use of effluent standards based upon concentration and water flow is less costly and more cost-effective in most instances for noncarcinogenic chronic cadmium risks than use of ambient or drinking water standards. This is true for mining and smelting up to a margin of safety of 9 for electroplating to a margin of safety of 5 and for phosphate fertilizer plants to a margin of 10. Control of drinking water costs slightly more and reaches the same degree of control in terms of margins of safety. Ambient controls cost the most because of loss of water use. Moreover, the polluter does not pay directly as in the case of effluent standards.

II. ''ZERO RISK'' (ALL PATHWAYS)

Table 31 extended Table 30 to include other exposure pathways, specifically ambient water (0.0004 ppb), ambient air (5 ppb), food (75 ppb) and smoking (40 ppb for 2 packs / day). When we factor in these pathways, water-borne cadmium is generally only a small proportion of the total exposure burden. This is true even for effluent from the three industries on low-flow rivers where significant exposure occurs (Table 31). The associated health effects are shown for both smokers and nonsmokers. The same type of analysis as for water pathways only is made except that the total burden from cadmium is included.

A. Effluent Standards

1. Mining and Smelting

If 120 ppb is believed to be the no effect level upper limit without a safety margin, it is possible, using effluent controls for mining and smelting, to reduce effluents below this level for all sources. This would mitigate risk for nonsmokers. However, the margin of safety would be decreased to a factor of 1.30 and 1.33 for successive controls (entries A1 and A2, Table 42). Smokers remain at risk at these levels although that risk is substantially reduced. We use factors of 10 and 1%, more to illustrate the impact of controls and cost-effectiveness than to realistically estimate actual kidney disease. One must essentially stop the industry to totally protect smokers.

a. Concentration at the Discharge Point

Entry A1 in Table 42 summarizes control applications. The marginal cost-effectiveness, based upon reduced smoker risk increases very rapidly. Incremental cost-effectiveness does so at a much lower rate. We have still not answered the key question, whether control of other sources would be more cost-effective. This type of question, which demands a multi-media approach to regulation, should, however, be asked by the regulator.

b. Concentration as a Function of Flow

Entry A2, Table 42, summarizes results similar to A1 except that here only a portion of the control expenditures are affected. The actual level achieved by closing the industries, resulting in a decontamination factor greater than 90, is 8 ppb higher than the previous case. This is because mines and smelters would still operate on average or higher flow rivers. The approach is more cost-effective in attaining the same results than the use of concentration alone.

Table 42
APPLICATION OF CONTROLS FOR CHRONIC CADMIUM HEALTH EFFECTS: ZERO RISK CONSIDERING ALL PATHWAYS

Entry number	Industry	Degree of reduction	Resultant water concentration	Total exposure Nonsmokers	Margin of safety (HE range)	Total exposure smokers	Margin of safety (HE range)	Cost of control × 1,000,000	Value of technology	Health effects reduced (best estimate) Nonsmokers	Smokers	Total	Cost effectiveness for water effluent control Incremental	Marginal
A1	Mining and smelting (concentration)	0	1030	1011	(50%)	1141	(55%)	0	0	0	0	0	0	0
		30	35	115	—	155	(10%)	1.3	0	9×10^4	3×10^4	12×10^4	11 — 220	11 — 220
		90	12	92	1.30	132	(1%)	2.6	0		3.4×10^4	12.4×10^4	21 — 420	325 — 6,500
		790	0	80	1.33	120	—	—	16×10^6		3.43×10^4	12.43×10^4	130 — 2600	$4.5 \times 10^4 - 9 \times 10^5$
A2	(Concentration) flow	0	1030	104	(50%)	1141	(55%)	0	0	0	0	0	0	0
		30	35	115	—	155	(10%)	0.39	0	9×10^4	3×10^4	12×10^4	3 — 65	3 — 65
		90	12	92	1.30	132	(1%)	0.75	0		3.4×10^4	12.4×10^4	6 — 120	90 — 1,800
		790	0	88	1.33	128	—	—	5×10^6		3.43×10^4	12.43×10^4	40 — 800	$1.4 \times 10^4 - 2.8 \times 10^5$
B1	Captive electroplating shops (concentration)	0	583	663	(40%)	703	(45%)	0	0	0	0	0	0	0
		14	42	122	—	162	(10%)	9.7×10^3	0	5×10^6	1×10^6	6×10^6	$1.6 \times 10^3 - 3.2 \times 10^4$	$1.6 \times 10^3 - 3.2 \times 10^4$
		33	18	98	1.20	138	(1%)	1.3×10^4	0		10.1×10^6	6.1×10^6	$2.1 \times 10^3 - 4.2 \times 10^4$	$3.3 \times 10^4 - 6.6 \times 10^5$
		100	6	86	1.40	126	—	1.9×10^4	5×10^6		1.11×10^6	6.11×10^6	$3.1 \times 10^3 - 6.2 \times 10^4$	$6.0 \times 10^5 - 1.2 \times 10^7$
B2	(Concentration and flow)	0	583	663	(40%)	703	(45%)	0	—	0	0	0	0	0
		14	42	122	—	162	(10%)	9.7×10^2		5×10^6	1×10^6	6×10^6	$1.6 \times 10^2 - 3.2 \times 10^3$	$1.6 \times 10^2 - 3.2 \times 10^3$
		33	18	98	1.20	138	(1%)	1.3×10^3			1.1×10^6	6.1×10^6	$2.1 \times 10^2 - 4.2 \times 10^3$	$3.3 \times 10^3 - 6.6 \times 10^4$
		100	6	86	1.40	126	—	1.9×10^3	—		1.11×10^6	6.11×10^6	$3.1 \times 10^2 - 6.2 \times 10^3$	$6.0 \times 10^4 - 1.2 \times 10^6$
C1	Phosphate fertilizer (concentration)	0	306	386	(30%)	426	(3%)	0	0	0	0	0	0	0
		30	10	90	1.3	130	(1%)	18		5×10^5	1.98×10^5	6.98×10^5	26 — 520	26 — 520
		730	0	80	1.5	120	—		3.7×10^9	5×10^5	2×10^5	7×10^5	$5.3 \times 10^4 - 1.1 \times 10^6$	$1.8 \times 10^6 - 3.7 \times 10^7$
C2	(Concentration and flow)	0	306	386	(30%)	426	(35%)	0	0					
		30	10	90	1.3	130	(1%)			5×10^5	1.98×10^5	6.98×10^5	2.60 — 52	2.60 — 52
		730	0	82	1.4	128	—	1.8	3.7×10^8	5×10^5	2×10^5	7×10^5	$5.3 \times 10^2 - 1.1 \times 10^4$	$1.8 \times 10^5 - 3.7 \times 10^6$

2. Electroplating Captive Shops

Entries B1 and B2, Table 42, summarize results. Adequate controls are available to maintain total intake below 120 ppb, under at least some conditions.

a. Concentration at the Discharge Point

This is summarized as entry B1, Table 42. Both incremental and marginal cost-effectiveness are very high.

b. Concentration as a Function of Flow

Entries appear in Table 42, B2. Cost-effectiveness improves by a factor of 10 since only 10% of the industry is controlled.

3. Phosphate Fertilizer Plants

Results are summarized in entries C1 and C2, Table 42. If a reduction factor greater than 30 is obtained to protect smokers, the industry would close. The value of technology lost is from Table 21.

a. Concentration at the Discharge Point

Entry C1, Table 47, shows that it is still relatively cost-effective to apply a first level of control at all phosphate fertilizer plants, but that both incremental and marginal costs will soar if the industry is closed.

b. Concentration as a Function of Flow

This is presented in entry C2. If plants on low-flow rivers are closed down, others will still add to the total.

B. Ambient Standards (Water Quality)

From Table 31, it is evident that smokers are already at the limit of the no effect level *wihtout* adding exposure from low-flow industry effluents. Nonsmokers have a margin of only 1.5. Ambient control of water quality to protect smokers must essentially close down all industries on low-flow rivers and still leave no margin of safety. Any water quality standard under 2 ppb prevents these industries from operating. A standard of 40 ppb allows some to operate with control, but does not provide a margin of safety for nonsmokers and possibly leads to increased kidney disease for smokers.

Since ambient water levels are such a small part of the total human exposure, ambient water standards cannot be set in the global context of controlling all risk to no effect levels.

C. Drinking Water Standards

Since drinking water provides only a small portion of dietary cadmium (approximately 75 ppm are ingested in food), control of cadmium in water at the global level does not make much sense.

III. AS LOW AS CAN BE MEASURED

The detection limit in water ranges from 0.005 to 2.5 mg/ℓ (ppb) depending upon method used.[137] Most field analytical measures are in the range of 0.2 to 0.7 ppm.

Measurements are not taken of chronic health effects from annual levels below 500 ppb intake. In addition, one can establish health impact relationships below 10 ppm/year only by models.

IV. AS LOW AS CAN BE CONTROLLED

As low as can be controlled is interpreted as the maximum control level achievable without shutdown. It amounts to picking the maximum control points in each situation. The standard is set at the level control would provide. Control can be applied by the total industry or for only those plants causing risk, i.e. those located on low-flow rivers.

For effluent controls from water, in Table 40, only the maximum entries are underlined in the column for required reduction. Compared with the first level of control, maximum control of mining and smelting and electroplating only buys an increased margin of safety. For phosphate fertilizer plants, the first level of control *is* the maximum level.

When considering all pathways, the effects of maximal control are underlined under the heading, Degree of Reduction, in Table 42. Maximal control of phosphate plants is the most cost-effective on a marginal basis, followed by mining and smelting. It is not, however, particularly cost-effective, on a marginal basis, for captive electroplating shops.

Entries E1 through E4, Table 41, are maximum control levels for drinking water.

V. AS LOW AS BACKGROUND

As low as background is interpreted in two ways:

1. Ambient background of 0.4 ppb for drinking water and 5.1 ppb for raw water
2. Total background of about 80 ppb including inhalation and ingestion (not including smokers)

The objective is to ensure that ambient levels are not significantly raised. Thus, one might set a standard by taking a small fraction of the natural variation in background. One possible criterion is the use of one standard deviation about the mean value of measurement.

Total cadmium in raw water has a mean of 5.1 ppb with standard deviation of 9.2, as derived from STORET data for the years 1968 to 1980. A water quality standard could be set upon this basis, i.e., 5.1 ppb ambient plus 9.2 ppb variation, or 15 ppb. If we assume a normal distribution for these measurements, about 15% of the sites will exceed the standard on a statistical basis alone. A standard of 9 to 10 ppb set above the local mean value, i.e., in addition to measured background, is an alternative approach. This standard could not be achieved by setting controls for mining and smelting on low-flow sites. One must apply maximal control for captive electroplating shops and phosphate fertilizer plants as indicated in column 4 (Resultant Water Concentration), Table 42. Comparing this standard to present EPA water quality standards (Table 43), the former is seen to exceed the total EPA ambient standard of 10 ppb.

The total ambient range, 80 ppb, is exclusive of smoking. A range of 60 to 190 ppb has been estimated[137] although a standard deviation has not been determined. If we assume standard deviation of 10 ppb from food and air and 10 ppb from water, the level is 20 ppb above measured mean levels. In this case, maximal control is necessary for mining and smelting, but plants on low-flow sites would not have to close to meet the standard. Only two control levels are required for captive electroplating shops, and phosphate fertilizer plants still have to provide controls.

Drinking water levels range from 0.2 to 0.4 ppb (Chapter 8) with a mean value near 0.4 ppb. The standard deviation cannot surpass 0.2 ppb which is significantly lower than the 10 ppb EPA standard. Drinking water standards of 1 ppb, 10 times more stringent than the present EPA level, could be set on this basis and not affect most existing supplies. This can be explained by the fact that raw surface water averages only 5.1 ppb and that cleanup during conversion to finished drinking water will remove cadmium.

Table 43
EPA WATER QUALITY STANDARDS

Cadmium

Recommendation[a]

	Recommendation[a]
Aquatic life	
Fresh water	
Hardness of 50 mg/ℓ	1.5 µg/ℓ
100 mg/ℓ	3 µg/ℓ
Salt water	
24-hr Average	4.5 µg/ℓ
Maximum at any time	59 µg/ℓ
Human health	10 µg/ℓ

Phenol

No aquatic life recommendations	
Human health	3.5 µg/ℓ

Note: Federal Register Notice of November 28, 1980. EPA Water Quality Criteria Recommendations[160] (state standards are generally based on these recommendations).

[a] µg/ℓ = 1 ppb.

Chapter 24

ACCEPTABLE RISK METHODS: CADMIUM NONCARCINOGENIC CHRONIC RISKS

It is only by risking our persons from one hour to another
that we live at all. And often enough our faith
beforehand in an uncertified result is the
only thing that makes the result
come true.

William James
The Will to Believe (1897)
Is Life Worth Living?

Risk levels are determined here by evaluating risks from our exposure model directly rather than by evaluating concentrations, amounts, or control capabilities. Thus, one need weigh only the health effect information summarized in Chapter 19. Such risks are usually expressed in probability of occurrence to an exposed individual; or, if the whole population is exposed, the probability to an "average" of the most sensitive members of the population. These risks are presented in the form of either annual or lifeime probability of acquiring a disease resulting in mortality and/or morbidity. Table 44 shows the noncarcinogenic chronic risks from industry cadmium effluents to low-flow rivers. The low-flow numbers are artifices used for chronic noncarcinogenic risks. For cancer, such risks are determined directly by the model. We obtain lifetime risk by multiplying these values by 70.

I. COMPARABLE RISK

One way to establish acceptable levels of chronic cadmium risk is by using comparable risks taken by members of society as benchmarks. In this case, risks must be involuntary (not in the control of those exposed) and result in disease and possible premature death. If such benchmarks exist, then the cadmium risk or even a small addition to it might be acceptable. While there are many such lists, one by Richard Wilson showing a number of items posing a one in a million risk level per year (Table 45), is particularly useful.[18]

In addition, as shown in Table 46, one can calculate overall risk levels for the U.S.[13] In this table, the last two columns refer to total years of life lost and years lost per year of exposure, respectively; the remaining columns are self-explanatory. Both methods are revealed preference approaches, but other methods could be used.

We can arbitrarily derive an acceptable risk by assuming that the pollutant does not add more than 1% to the risk of drinking water in, for example, Miami (see Table 45). This results in a risk level of 10^{-8}/year with an odorless concentration level of 80 ppb (Table 44). A standard at this level requires only first level control to achieve acceptable effluent control. Moreover, chronic cadmium risks are basically insensitive to levels below 100/year since all first level controls result in risk well below 10^{-8}/year. Risks of 0.01/year or higher are unquestionably unacceptable. Application of margins of safety reduces amounts to 40 ppb and 8 ppb for levels of 2 and 10, respectively.

A margin of about two does not cause further control expenditure. Drinking water and ambient levels are already below any chronic noncarcinogenic risk level.

In summary, risk comparison has little meaning when, as with cadmium, there are easily attained no effect levels and risk is far above these levels.

Table 44
SOME ESTIMATED RISK LEVELS FOR CHRONIC CADMIUM EXPOSURE

Concentration (ppb)	Percent affected by kidney disease	Annual risk to exposed population	
		Best estimate	Low range
1100	50	5×10^{-1}/year	5×10^{-2}/year
660	40	4×10^{-1}/year	4×10^{-2}/year
300	30	3×10^{-1}/year	3×10^{-2}/year
150	10	1×10^{-1}/year	1×10^{-2}/year
130	1[a]	1×10^{-2}/year	1×10^{-3}/year
100	0.001[a]	1×10^{-5}/year	1×10^{-6}/year
80	0.000001[a]	1×10^{-8}/year	1×10^{-9}/year
10	—	—	—

[a] These numbers are hypothesized to illustrate that there may be members of the exposed population sensitive to cadmium intake levels below 120 ppb.

Table 45
RISKS WHICH INCREASE CHANCE OF DEATH BY 0.000001[a]

Smoking 1.4 cigaretes	Cancer, heart disease
Drinking 1/2 ℓ of wine	Cirrhosis of the liver
Spending 1 hr in a coal mine	Black lung disease
Spending 3 hr in a coal mine	Accident
Living 2 days in New York or Boston	Air pollution
Traveling 6 min by canoe	Accident
Traveling 10 mi by bicycle	Accident
Traveling 300 mi by car	Accident
Flying 1000 mi by jet	Accident
Flying 6000 mi by jet	Cancer caused by cosmic radiation
Living 2 months in Denver on vacation from New York	Cancer caused by cosmic radiation
Living 2 months in average stone or brick building	Cancer caused by natural radioactivity
1 chest X-ray taken in a good hospital	Cancer caused by radiation
Living 2 months with a cigarette smoker	Cancer, heart disease
Eating 40 tablespoons of peanut butter	Liver cancer caused by aflatoxin B
Drinking Miami drinking water for 1 year	Cancer caused by chloroform
Drinking 30 12-oz cans of diet soda	Cancer caused by saccharin
Living 5 years at site boundary of a typical nuclear power plant in the open	Cancer caused by radiation
Drinking 1000 24-oz soft drinks from recently banned plastic bottles	Cancer from acrylonitrile monomer
Living 20 years near PVC plant	Cancer caused by vinyl chloride (1976 standard)
Living 150 years within 20 miles of a nuclear power plant	Cancer caused by radiation
Eating 100 charcoal broiled steaks	Cancer from benzopyrene
Risk of accident by living within 5 mi of a nuclear reactor for 50 years	Cancer caused by radiation

[a] 1 Part in 1 million

Table 46

SUMMARY OF ABSOLUTE RISKS LEVELS FOR DIFFERENT TYPES OF RISK

Type of risk		Class of consequence				
Index	Description	fat/year (1)	he/year (2)	$/year (3)	year (5)	year/year (6)
	Naturally occurring					
1	Thresholds of concern (I)	1×10^{-10}			3×10^{-7}	4×10^{-9}
2	Catastrophic risk (I)	1×10^{-6}	5×10^{-6}	0.2	3×10^{-3}	4×10^{-5}
3	Ordinary risk (I)	7×10^{-5}	4×10^{-4}	3	0.2	2×10^{-3}
	Man-originated					
4	Thresholds of concern (I)	1×10^{-11}			3×10^{-8}	4×10^{-10}
5	Catastrophic		5×10^{-7}	2×10^{-2}	3×10^{-4}	4×10^{-6}
	Involuntary (I)	1×10^{-7}	2×10^{-6}	0.4	6×10^{-3}	8×10^{-5}
	Voluntary (V)	2×10^{-6}	3×10^{-6}	0.4	6×10^{-2}	8×10^{-4}
	Regulated voluntary (RV)	3×10^{-5}				
6	Ordinary		3×10^{-5}	1	1×10^{-2}	2×10^{-4}
	Involuntary (I)	5×10^{-6}	3×10^{-1}	200	1	0.1
	Voluntary (V)	6×10^{-4}	6×10^{-2}	30	0.1	1×10^{-2}
	Regulated voluntary (RV)	1×10^{-4}				
	Man-triggered					
7	Catastrophic		1×10^{-6}	4×10^{-2}	6×10^{-4}	8×10^{-6}
	Involuntary (I)	2×10^{-7}	4×10^{-6}	0.8	6×10^{-3}	2×10^{-4}
	Voluntary (V)	4×10^{-6}				
8	Ordinary				3×10^{-2}	4×10^{-4}
	Involuntary (I)	1×10^{-5}			2	0.2
	Voluntary (V)	1×10^{-3}			0.2	2×10^{-2}
	Regulated voluntary (RV)	2×10^{-4}				

Note: he = health effects

II. ARBITRARY RISK NUMBERS

If one assumes an arbitrary acceptable risk level, such as 10^{-5}/year, 10^{-6}/year, or 10^{-7}/year, or some other such fixed acceptable risk for the general population, we can then derive a standard directly from the exposure-risk model. The uncertainties in this model may, however, be very large. A 10^{-5} level results in 100 ppb. Any lower level would probably result in 80 ppb since computations in this range are totally speculative and uncertain. Margins of safety would further reduce these numbers. For example, with a margin of safety with a factor of 2, we need no further controls at any set level between 10^{-5}/year and 10^{-9}/year. Note that this method is insensitive for the chronic noncarcinogenic cadmium risk.

III. A SET VALUE FOR DOLLARS TO BE SPENT

The objective here is to limit spending for control to a fixed amount. One might establish such a level in a variety of ways. These are

- Actual resources available
- Resource limits established by governmental budgeting processes
- A level just below which an industry might consider closing down
- Limit on taxpayer burdens
- A percentage of the value of the technology
- Other approaches

All approaches are either specific to a situation or arbitrary. Because of availability, for this analysis we will use the percentage of value of technology, selecting arbitrary figures of 1% and 10%.

Results for effluent controls appear in Table 47. In all cases, except those for phosphate fertilizers, expenditures at any level of control exceed 1% of the value of technology. At 10% of the total value of technology for mining and smelting, only a first level of control is applied. Any controls at captive electroplating shops exceed the value of technology, even at 100%.

On this basis, the first level of control at mines and smelters costs about 8% of the value of technology. If all mines and smelter effluents were held to this level by a concentration standard at the discharge point, expenses would be passed on to users. If only those on low-flow rivers were affected, the two plants would have a higher, possibly uncompetitive, cost structure. Exposure is at a level where health effects are well below the no observable effect level, but only with a margin of three. Further control to achieve a higher margin of safety would shut down affected plants.

Captive electroplating shops cannot be economically controlled by cadmium effluent reduction alone. If other processes remove other pollutants and this, to some extent, reduces the amount of cadmium, it may be effective. However, we base these conclusions upon industry averages which have great uncertainty. It may well be that certain large and efficient plants could afford controls while marginal operators would be forced out of the market.

An alternative for captive electroplating shops is to prohibit use of the water on low-flow rivers by using ambient water quality standards. Water would have to be obtained elsewhere and captive electroplating shops on low-flow rivers would be subsidized. This would cost $280 million/year. A ridiculous expenditure that is 35% higher than the total value of the industry.

There are no major problems in setting one level of control for phosphate fertilizer plants. Control expenditures represent only 0.04% of the value of technology and provide a margin of safety of 3.

IV. A SET VALUE FOR RISK REDUCTION

Set values for risk reduction can be established by addressing either the level of reduced risk or residual risk after reduction. In the first case, reduction may be a level of health effect avoided, a fraction of total risk, or some other functional relationship. The residual situation is similar to the comparison and arbitrary cases already addressed.

If one wishes to reduce risks from the highest uncontrolled level by some factor, perhaps 10 or 100, there must be a basis for establishing that factor. It might be a breakpoint in the dose-effect curve, acceptance of a linear dose-effect model which implies some risk at any level of exposure above zero, or some other judgment.

It may not be feasible to consider relative reduction in the form of a reduction factor because of the large uncertainty in the absolute risk estimates. One does not usually know the actual level of risk to be reduced. This is, however, not as great a problem in the relative reduction case.

There are additional limitations in the highly nonlinear situation for chronic risks. From Table 44, it is evident that a factor of 5 for reduction occurs at 150 ppb, a factor of 10 at 145 ppb, a factor of 50 at 130 ppb, and perhaps a factor of 100 at 125 ppb. In all instances, the first level of control results in a risk reduction factor of more than 100. Without controls, there is no risk reduction. There is no in-between. On the other hand, margins of safety are not built in since some risk is accepted and only first level of control is required in any situation.

Table 47
EFFLUENT CONTROL BASED UPON PERCENTAGES OF VALUE OF TECHNOLOGY

Industry	Value of technology ($ × 10^6)	Plants affected	Control reduction factor	1% of Value of technology ($ × 10^6)	Control cost ($ × 10^6)	10% of Value of technology ($ × 10^6)	Number of jobs affected by shutdown
Mining and smelting							
Conc std	16	7	30	(0.16)[a]	1.3	1.6	4,600
			90	(0.16)	2.6	(1.6)	
Conc + flow	5	2	30	(0.05)	0.39	0.50	460
			90	(0.05)	0.75	(0.50)	
Electroplating							
Captive shops							
Conc. std.	6,050[b]	6,050	6	(61)	9,700	(610)	40,000
			12	(61)	9,700	(610)	
			30	(61)	13,000	(610)	
			60	(61)	19,000	(610)	
Conc + flow	605	605	6	(6.1)	970	(61)	4,000
			12	(6.1)	970	(61)	
			30	(6.1)	1,300	(61)	
			60	(6.1)	1,900	(91)	
Ambient water quality					820		
Phosphate fertilizer plants							
Conc std	3,700	92	30	37	18	370	16,000
Conc + flow	370	9	30	3.7	1.8	37	1,600

a ( ) Indicates that control costs exceed the value of technology at the fractional level shown.

b Based upon the number of plants of each type, assuming equal value from Table 21.

Chapter 25

ECONOMIC AND DESIGN APPROACHES: CADMIUM NONCARCINOGENIC CHRONIC RISKS

Mere parsimony is not economy. . . . Expense,
and great expense, may be an essential
part of true economy.

Economy is a distributive virtue, and consists
not in saving but selection. Parsimony
requires no providence, no sagacity,
no powers of combination, no
comparison, no judgment.

Edmund Burke (1729—1797)
Letter to a Noble Lord (1796)

When weighing risk factors to achieve maximum protection *with* minimum adverse economic impact, one must indeed employ sagacity and judgment, as well as the ability to combine and compare diverse elements. In this chapter, we will discuss both economic and design approaches to standard setting to mitigate chronic cadmium risks.

I. ECONOMIC APPROACHES

Although the only indirect consideration in these approaches is the loss of jobs (Table 47), other indirect and intangible cost and values may also be significantly affected.

A. Marginal Cost of Risk Reduction

There is a range of 20 between the best estimate and a possible high range for marginal cost of risk reduction. At what cost-effectiveness should we set a marginal cost cut-off point?* Many studies have explored the problem of setting a value for this criterion.[13] These values range between $50,000 to $1 million per health effect avoided. Any costs must be tied to a base year to prevent inflation from causing a value to exceed a limit. The rulemaking for radiation protection levels for nuclear power plants uses a rule of thumb of $100,000 to $500,000 per health effect to determine effectiveness of standards.[169] We have evaluated three levels, $100,000; $250,000; and $500,000 per health effect avoided. Again, increased spending for margins of safety that do not change health impact has infinitely high cost-effectiveness.

When we evaluate only water pathways (Tables 40 and 41), all first-level control methods are cost-effective at even the high range of the estimate. The highest amount is $97,000 per health effect avoided. This is for the high range estimate for concentration standards at captive electroplating shops. All second-level controls are cost-ineffective. Thus, a criterion for any method of standard setting includes applying first-level controls.

When one assesses all pathways (Table 31), the situation changes. Using best estimate values, our results are tabulated in Table 48 (high-range values in parentheses). In general, first-level controls for mining and smelting and phosphate fertilizers are very cost-effective, even at the high range. Even industry shut down is cost-effective at the best estimate level, but only in one instance at the high-range level. For electroplating, the range of criteria changes the level of control and the high range reduces control level one step.

---

* Often interpreted as a value of a life saved.

Table 48
RESULTS OF COST-EFFECTIVE ANALYSIS WITH
BEST ESTIMATE VALUES[a]

Industry	Low range (high range)[b]		
	$100,000/he	$250,000/he	$500,000/he
Mining and smelting			
Concentration	SD[c] (90)	SD (90)	SD (90)
Concentration + flow	SD (90)	SD (90)	SD (SD)
Captive electroplating			
Concentration	33 (14)	33 (14)	100 (14)
Concentration + flow	100 (33)	100 (33)	100 (33)
Phosphate fertilizer			
Concentration	30 (30)	30 (30)	30 (30)
Concentration + flow	30 (30)	SD (30)	SD (30)

Note: he = Health effects.

[a] Degree of reduction factor required per Table 42.
[b] ( ) = High range.
[c] SD = shutdown of industry.

The key issue is the number of health effects. Even at the high range, control is cost-effective for captive electroplating shops at the first level, but attaining it effectively puts these companies out of business. This is a situation where cost-effectiveness of reduction is insufficient in itself to establish levels of standards.*

1. Tax Incentives

A pollution tax for low-flow plants is an alternative to controls. There are a least two ways to determine tax level. The first is to set tax at the same, or slightly higher, cost equivalency as the control needed to reach an effective level. The objective is to motivate the industry to invest in controls, yet allow it to time installation to take advantage of economic trade-offs among investment capital, current expense, equipment depreciation, and other tax incentives. An alternate approach is to use the value of health effects reduced. One would use a cost equivalent for life shortening. This assumes ambient water and drinking water controls protect health. Here the cost of government implementation is reimbursed equitably through taxes to polluting industries, based upon both the expenses of developing these standards (they have higher costs than effluent standards — see Table 41) and the risk burden as determined by the river model. This approach may be favored for setting finished drinking water standards after implementing controls for other pollutants, (indirectly reducing the cadmium levels). In this situation, a pollution tax on captive electroplating shops might be effective.

These standards might more easily be based upon the amount rather than concentration discharged during low-flow conditions. Regardless of the method chosen, one must base equitable allocation either upon the contribution of the effluent of a plant to health impact or upon the initial cost to someone other than the polluter. In the first instance, there is no assurance that a tax will avoid health effects. In the second, action is taken to avoid health

* On the other hand, acknowledging problems of uncertainty in unrelated risk estimates, we may still use the cost-effective levels to determine whether control is more cost-effective in air, food, or other media. For a complete assessment, we must ascertain both the control methods and their cost and effectiveness. This, however, was outside the scope of our analysis.

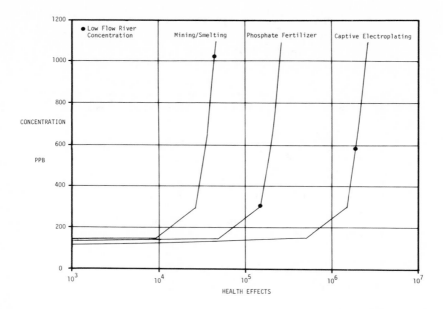

FIGURE 27.   Nonlinearity of concentration vs. health effects (for each indusry of cancer).

effects and then to reallocate costs to the polluter. Whether this can be done fairly remains to be seen.

Figure 27, based upon the exposure model, illustrates the nonlinearity of concentration vs. health effects for each of the three industries of concern. One can convert concentration to amount per unit river flow to provide a measurable volume release. Tax levels may then be based on either amounts released or amount released plus unit river flow. The former taxes all plants independent of flow, the latter only plants on low-flow rivers.

The value for each industry affected is set at the no control points indicated by dots on Figure 27. The tax basis, assuming no control, is shown in Table 49. For health effect equivalent, columns 6 through 8, the tax basis exceeds the value of technology by at least 1 order of magnitude. Taxes predicated upon cost of effluent controls, column 3, are effective for the mining and smelting and the phosphate fertilizer industries, but not for electroplating. Even where this strategy is effective, pollution and resultant adverse health effects might occur. However, taxes are collected from all plants on an amount basis alone, an equitable arrangement. Prorating the expense of either cleaning up drinking water by taxes (column 5) or of alternate water supplies (column 4) depends upon who pays, the entire industry or just plants on low-flow rivers. Actually, these approaches are useful only when applied to captive electroplating shops which would be subject to taxation regardless of water flow. Drinking water control taxes prorate to about 23% of the value of technology, and taxes to pay the cost of alternate water supplies to about 14%. It is the only approach in which the total industry pays that would not, in our estimation, cause some plants to choose to go out of business. All other taxation methods provide few incentives over more direct control methods.

We emphasize that the tax incentive approaches shown are only used to evaluate feasibility. We have not addressed certain problems in using them, including tax jurisdiction, legality, and equitable redistribution of funds.

2. Risk/Cost/Benefit Balancing

As discussed previously, risk/cost/benefit balancing theoretically involves finding the equivalent level between the marginal cost of risk reduction and the marginal cost of industry

Table 49

POSSIBLE BASIS FOR EFFLUENT TAXES FOR NO CONTROL (PER PLANT)

Industry	Number of plants taxed	Control equivalent (per plant)			Health effect equivalent (per plant)			Average value of technology per plant
		Effluent control	Ambient level of use	Drinking water	$100 he	$250 K/he	$500 K/he	
(1)	(2)	(3)	(4)	(5)	(6)	(7)	(8)	(9)
Mining and smelting								
Amount only	7	1.9×10^{5a}	2.9×10^6	5.0×10^{5a}	1.8×10^9	4.5×10^9	9.0×10^9	2.3×10^6
Amount + flow	2	1.9×10^{5a}	1.0×10^7	1.8×10^{6a}	6.3×10^9	1.6×10^{10}	3.2×10^{10}	2.3×10^6
Captive electroplating								
Amount only	6050	1.6×10^6	1.4×10^{5a}	2.3×10^{5a}	3.3×10^7	8.3×10^7	1.7×10^8	1.0×10^6
Amount + flow	605	1.6×10^6	1.4×10^6	2.3×10^6	2.3×10^8	8.3×10^8	1.7×10^9	1.0×10^6
Phosphate fertilizer								
Amount only	92	2.0×10^{5a}	2.9×10^{6a}	5.0×10^{5a}	3.6×10^8	9.0×10^8	1.8×10^9	4.0×10^7
Amount + flow	9	2.0×10^{5a}	2.9×10^{7a}	5.0×10^{6a}	3.6×10^9	9.0×10^9	1.8×10^{10}	4.0×10^7

Note: he = health effect.

a Cost is less than value of technology.

benefit production when risk and benefits are measured on the same scale. Within limits of uncertainty, the marginal cost of risk reduction has already been ascertained. In principle, one would like to obtain the marginal benefit of industry production in terms of the direct value of technology, employment, and indirect benefits to society, etc., but these items are on different scales than risk or health-effects. Thus, conversion into a common scale is necessary for quantitative assessment (this may not be the case of qualitative assessments). One such scale is dollars, the use of which suffers the difficulties of converting all parameters to monetary units.

Using dollars, the marginal risk reduction curve can be established in terms of control costs expended to reduce the equivalent dollar value of health effects reduced on a marginal basis. This may be accomplished by converting health effects to dollars as was considered for the marginal cost of risk reduction. The value of technology, however, cannot be ascertained directly as dollars are spent to improve the value of technology, i.e., increased investment in production facilities as opposed to pollution abatement activities. Even if increased employment could be converted into dollars, the indirect influences of increased production cannot be easily evaluated. The best that can be done here is to evaluate the loss of value of technology by increased control spending. On this basis, the marginal cost-effectiveness of value of technology loss is linear with control cost up to the level of shutdown. We assume here that a dollar spent on control must be made up by an equivalent value of technology. This is, of course, an over-simplification since fixed and variable costs, as well as break-even analysis, are necessary on an individual plant basis.

We have used only the best estimate of health effects and a value of $100,000 per effect for the marginal cost-effectiveness of risk reduction. For the value of technology lost, we assume indirect and employment costs contribute a factor of 10 over the direct value of technology. Figure 28 illustrates the situation with respect to the mining and smelting industry considering the total ambient situation as shown in entry A2, Table 42. At first glance, it appears that one can easily determine values of equal slopes, but this is not so since the left ordinate scale is actually three orders of magnitude larger than the right. Even using the lower health effect estimate and a factor of 100 for technology indirect value, the scales differ by a factor of 10. Thus, resolution of the method is so poor as to be of no use for this or the remaining two industries.

B. Design Approaches

Control design methods or classes of design methods can be specified in standards. The Clean Water Act calls for two such classes: Best Practical Control (BPT) and Best Available Control (BAT).

These control design methods can be applied either to the entire industry or to only that portion on low-flow river sites, depending upon the "trigger" mechanism to effect control, i.e,, whether based upon river flow.

1. Best Practical Control Technology

For the mining and smelting and phosphate fertilizer industries, BPT requires only first-level control. Controls would be applied for captive electroplating shops only when shut-down is not implied. The choice is either no control or shut-down. There is no middle ground.

The EPA states that BPT "Shall include consideration of the total cost of the application of technology in relation to effluent reduction benefits to be achieved from such applications and shall also take into account the age of equipment and facilities involved, the process employed, engineering aspects of the application of various types of control technology, process changes, non-water quality environmental impacts (including energy requirements) and any other factors the Administration deems applicable."[170]

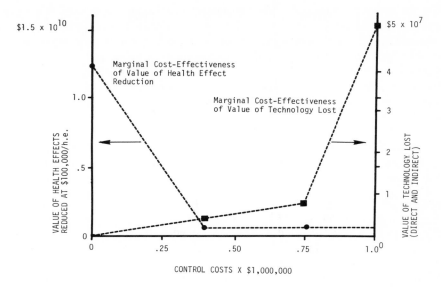

FIGURE 28.   Risk/cost/benefit balancing at the margin for mining and smelting effluent controls.

2. Best Available Control Technology

The technology implies a second level of control for mining and smelting plants, essentially improving the margin of safety. It would not affect the phosphate fertilizer industry since the effect of BAT here is identical to that of BPT. The captive elecroplating shops must apply three levels of control, closing down the industry.

The EPA defines BAT as: ''The Best Available Control Technology economically achievable, which will result in reasonable further progress toward the national goal of eliminating the discharge of all pollutants. In general, this technology represents at a minimum, the very best economically achievable performance in any industrial category or subcategory.''

Chapter 26

SUMMARY — CADMIUM NONCARCINOGENIC CHRONIC RISKS

*The inherent vice of capitalism is the
unequal sharing of blessings; the
inherent virtue of socialism is the
equal sharing of miseries.*

Saying
Winston Churchill (1874—1965)

Equity in the sharing of risks is difficult to achieve under any political system. However, it is a factor a government responsible to the people must weigh in making all its decisions. It has been one of the factors included in our assessment of the approaches to setting standards for noncarcinogenic chronic cadmium risks discussed individually in the three preceding chapters and summarized here.

Of those approaches, a zero risk criterion is reasonable for these risks because the no effect level is, in general, higher than the total ambient concentration. For water pathways alone, zero risk is feasible for both mining and smelting and phosphate fertilizer industries. The major issue in these cases is the degree of margin of safety and the required expenditures. For effluent standards, it is whether to regulate the entire industry or only low-flow sites. However, costs for the captive electroplating shops, if they are to achieve any degree of control, will exceed the value of technology.

In the latter instance, one can set ambient water quality standards or drinking water standards and controls, but the water user, not the polluter, must pay. One way of ameliorating this inequity is to impose an effluent tax forcing the polluter to foot the bill for controls. As was seen in Table 49, this is economically feasible only if the total industry is taxed regardless of their actual contribution to deleterious health effects.

None of the other techniques such as adopting a margin of safety, determining the extent and degree of control, or predicting shut-down for electroplating captive shops contribute very much more to resolving issues. Because the margin of safety is all that is bought by further control, the marginal cost of risk reduction does not help, except in determining total ambient risk. Risk/cost/benefit balancing does not work at all in these cases. Best practical and best available technology helps define degree of control for mining and smelting but not for the other industries. As low as can be measured and as low as background methods add nothing to the solution. As low as can be controlled and BAT are essentially the same in this instance.

The methods and types of standards are both industry- and situation-specific. Methods which work in all three situations result in some level of industry closure.

Note that calculations for the electroplating industry use average figures and several simplifying assumptions. For example, many plants are, of course, smaller or larger than the average; they may or may not need the first two stages of control for chromium reduction and cyanide oxidation; and we have not addressed the issue of co-location. As a result, our analysis may not typify the industry, but does illustrate the problem of small industrial operations with high control costs.

Another alternative, controlling discharge of wastes to municipal treatment plants, has been pondered by EPA. However, this can only be used for small volume dischargers such as those in the electroplating industry. Discharge volumes for the mining and smelting and the phosphate fertilizer plants are too large for this approach (Table 12).

This exercise shows that for chronic noncarcinogenic risks of large consequence, both

---

EPA AND ELECTROPLATING

It is interesting that EPA regulations for electroplating plants, proposed at the time of this writing, if implemented, are expected to close 20% of the industry. Expenses entailed are based upon many pollutants, but there is a marked correspondence between the river exposure model developed here and EPA's models. Although independently derived, the results are in the same range.

---

quantitative and qualitative methods are inadequate in providing answers when control costs can cause operations to close their doors. In this instance, not only are marginal operators affected, but the entire industry. The decision must, therefore, be made on a purely political basis, balancing a viable industry and its direct and indirect value against potential health impact. Using substitutes provides an alternative to an approach based on value of the product, but results in transfer of value of technology from one industry to another. This again requires a political decision. Quantitative analysis can determine if substitution preserves the value of technology and reduces health impact, but can address neither the inequities nor the temporary and permanent translocation problems.

Part VIII

EVALUATING APPROACHES TO STANDARDS FOR CADMIUM CANCER RISK

Health and intellect are the two blessings of life.

Menander (c. 342—292 B.C.)
Monostikoi (Single Lines)

Evaluating methods of setting risk levels for standards protecting against cancer risk necessitates a different approach than that used when protecting against other chronic risk. We use a linear dose-effect relationship for cancer (although, as discussed in Part IV, other models are possible). Zero is at the origin of the relationship. Only the slope changes as we express the uncertainty in potency. While other models may add dose rate relationships to provide nonlinearity at very low dose, this model shows the impact of standard setting in strictly linear situations.

We use three levels of potency, maximum, mid-range, and zero level. The maximum provides a larger margin of safety than does the mid-range estimate. Although we have not used a formal method of providing confidence levels for extrapolation of dose-effect models, they are well within the imprecision of the total model.

One implication of linearity is that a dose reduction at any particular site is independent of any other. Thus, only local considerations are necessary, and we can directly scale dose-effects. On this basis, each industry is addressed separately to determine its contribution to water pollution. That contribution will later be compared with other risks and background levels.*

Cost-effectiveness is based upon annual costs of control and reduction in annual health impact. Cancer impact is presented on a lifetime basis. Thus, values from Chapter 13 must be divided by 70 (used as the average length of a life) to obtain an annual average health impact. These values, as well as cancer risk estimate for low flow cases, are shown in Table 50.

---

* In many cases, health effects are shown to three significant figures. Our purpose in so doing is to illustrate calculation of marginal cost-effectiveness, not to ascribe a level of precision to the estimates.

Table 50
EFFLUENT CONTROL APPLIED TO CADMIUM FOR CANCER IMPACT

Industry (number of plants)	Pipe level concentration	Downstream concentration	Reduction factor	Incremental cost ($ million)	Incremental value of technology ($ million)	Health effects reduced[a] (unreduced) Maximum	Mid-range	Minimum	Marginal cost of risk reduction Maximum ($/he)	Mid-range ($/he)	Minimum ($/he)	% of Total cancer impact
A1 Mining and smelting												
Average Flow (7)												
Level 1	15 ppm	8 ppb	0	0	(16)	(0.1)	(0.001)	0				
2	500 ppb	0.3	30	1.3		0.097	0.00097	0	1.3×10^7	1.3×10^9		
3	200	0.1	90	1.3	13.4	0.002	0.00002	0	6.5×10^8	6.5×10^{10}		
	0	0				0.001	0.00001	0	1.3×10^{10}	1.3×10^{12}		
A2 Low Flow (2)	15 ppm	1031 ppb	0	0	(5)	(3.5)	(0.035)	0				9
Level 1		35	30	0.39		3.38	0.0338	0	1.2×10^5	1.2×10^7		
2		12	90	0.36	4.25	0.08	0.0008	0	4.5×10^6	4.5×10^8		
3		0				0.04	0.0004	0	1.1×10^8	1.1×10^{10}		
B1 Job Shops												
Average Flow (3000)	3.1 ppm	0.06 ppb	0	0	(3000)	(0.03)	$(0.3) \times 10^{-3}$	0				90
Level 1	220 ppb	0.004	14	680		0.0278	0.278	0	2.4×10^{10}	2.4×10^{12}		
2	94	0.002	33	60		0.0013	0.013	0	4.6×10^{10}	4.6×10^{12}		
3	3	0.0006	100	140	2120	0.0006	0.006	0	2.3×10^{11}	2.3×10^{13}		
4	0	0				0.0003	0.003	0	7.1×10^{12}	7.1×10^{14}		
B2 Low Flow (300)	3.1 ppm	8.5 ppb	0	0	(300)	(0.4)	(0.004)	0				<.02
Level 1		0.61	14	68		0.371	0.00371	0	1.8×10^8	1.8×10^{10}		
2		0.26	33	6		0.017	0.00017	0	3.5×10^8	3.5×10^{10}		
3		0.09	100	14	212	0.008	0.00008	0	1.8×10^9	1.8×10^{11}		
4		0				0.004×10^{-3b}	0.00004×10^{-5b}	0	5.3×10^{10}	5.3×10^{12}		
C1 Pb Manufacturers												
Average flow (400)	3.1 ppm	0.004 ppb	0	0	(400)	(3)	(3)	0				10
Level 1	220 ppb	0.00030	14	16		2.79	2.79	0	5.7×10^9	5.7×10^{11}		
2	94	0.00012	33	3		0.12	0.12	0	2.5×10^{10}	2.5×10^{12}		
3	31	0.00004	100	10	320	0.06	0.06	0	1.7×10^{11}	1.7×10^{13}		
4	0	0				0.03×10^{-3b}	0.03×10^{-5b}	0	1.1×10^{13}	1.1×10^{15}		
C2 Low flow (40)	3.1 ppm	0.06 ppb	0	0	(40)	(3)	(3)	0				
Level 1		0.0040	14	1.6		2.79	2.79	0	5.7×10^8	5.7×10^{10}		
2		0.0020	33	0.3		0.12	0.12	0	2.5×10^9	2.5×10^{11}		
3		0.0006	100	1.0	32	0.06	0.06	0	1.7×10^{10}	1.7×10^{12}		
4		0				0.03×10^{-3b}	0.03×10^{-5b}	0	1.1×10^{12}	1.1×10^{14}		

											he
D1 Captive shops											
Average flow (6050)											
Level 1	15 ppm	4.5 ppb	0	0	(6050)	(1)	(0.01)	0	4.2×10^{9}	4.2×10^{11}	
2	1.071	0.321	3900	14		0.93	0.0093	0	1.8×10^{10}	1.8×10^{12}	
3	0.455	0.136	730	33		0.04	0.0004	0	4.7×10^{10}	4.7×10^{12}	
	8	0				0.03	0.0003	0			
D2 Low flow (605)											
Level 1	15 ppm	583 ppb	0	0	1420^c	(30)	(0.3)	0	1.4×10^{7}	1.4×10^{9}	
2		41.63	390	14	(605)	27.9	0.279	0	6.1×10^{7}	6.1×10^{7}	
3		17.67	73	33	142^c	1.2	0.012	0	1.6×10^{8}	1.6×10^{10}	
		0				0.9	0.009	0			
Pigments											
Average flow (6)											
Level 1	1.3 ipm	0.002 ppb	0	0	(163)	(2) $\times 10^{-5}$	(2) $\times 10^{-7}$	0			<0.02
E1 Level 1	0.002	3.5×10^{-6}	571	0.23	162	1.996	1.996	0	1.2×10^{10}	1.2×10^{12}	
2	0.001	2.0×10^{-6}	1,000	0.02		0.002	0.002	0	1.0×10^{12}	1.0×10^{14}	
3	0.0006	1.0×10^{-6}	2,000	0.12		0.001	0.001	0	1.2×10^{13}	1.2×10^{15}	
4	0	0				0.001	0.001	0	1.6×10^{16}	1.6×10^{18}	
Low flow (2)	1.3 ppm	0.2 ppb	0	0.08	(54)	(6) $\times 10^{-4}$	(6) $\times 10^{-6}$	0			
E2 Level 1		3.5×10^{-4}	571	0.08		5.988	5.988	0	1.3×10^{8}	1.3×10^{10}	
2		2.0×10^{-6}	1,000	0.01		0.006	0.006	0	1.7×10^{10}	1.7×10^{12}	
3		1.0×10^{-6}	2,000	0.04		0.003	0.003	0	1.3×10^{11}	1.3×10^{13}	
4		0				0.003	0.003	0	1.8×10^{14}	1.8×10^{16}	
F1 Plastics (193)	0.02 ppm	0.0007 ppb			53.72	$\ll 1$	$\ll 1$	0			0.25
F2 Ni-Cd batteries (10)	0.77	0.010			(30)	$\ll 1$	$\ll 1$	0			<0.02^c
F3 Iron and steel (136)	0.07	0.0026			(104)	$\ll 1$	$\ll 1$	0			2^c
Phosphate fertilizer					(42,000)			0			5^c
G1 Average flow (92)											
Level 1	1.6 ppm	2.3 ppb	0	0	(3,700)	(0.4)	(0.004)	0	4.6×10^{7}	4.6×10^{9}	3
2	0.5	0.08	30	18	3,682	0.39	0.0039	0	3.7×10^{11}	3.7×10^{13}	
	0	0			(370)	0.01	0.0001	0			
G2 Low flow (9)											
Level 1	16.0 ppm	306 ppb	0	0	368	(5)	(0.05)	0	4.5×10^{5}	4.5×10^{7}	79
2		10	30	2		4.8	0.048	0	1.8×10^{9}	1.8×10^{11}	
		0				0.2	0.002	0			

Note: he = Health effects.

a Carried to three significant figures only to show effects. This precision should not be assumed to imply that information is meaningful.

b Indicates order of magnitude for the following category.

c Low-flow cases.

Chapter 27

RISK AVERSION METHODS — CADMIUM CANCER RISK

Avoid the reeking herd,
Shun the polluted flock,
Live like that stoic bird
The eagle of the rock

Elinor Hoyt Wylie
The Eagle and the Mole (1921)

Unfortunately, some things, such as exposure to relatively long-lived and ubiquitous toxic substances, may be independent of personal decisions as to where and how to live. In discussing cancer risk aversion approaches to evaluate cadmium standards, one must remember that cadmium has these properties.

I. ZERO RISK (WATER PATHWAYS ONLY)

For a nonthreshold pollutant, zero risk can be achieved in at least three different ways:

1.   Using controls that prevent any discharge
2.   Substituting a zero risk alternative
3.   Shutting down the industry

In any of the industries viewed, it is certainly possible to use permanent holding ponds and evaporation at very high cost, but this may increase environmental risks from polluted land and land fills, with redirection to other, possibly more risky, pathways. Shutdown of these industries would indeed be possible and in one industry guaranteed.

Some costs of adopting this approach are not readily apparent since the value of technology measures only the direct impact, and the indirect impact may even be greater. Use of available substitutes in some cases entails loss of advantages peculiar to the product (and not to the substitute) with resultant long-term societal losses. Short-terms losses occur from industry translocation. If one chooses zero risk, this extreme condition must be faced.

A. Effluent Standards

Table 56 summarizes available effluent controls mitigating cancer risk from cadmium industries. The water emissions of four industries result in exposure levels so low that risk measurement is virtually impossible. Nevertheless, we have investigated in detail one of these industries, pigments, as an example.

1. Mining and Smelting

The only way of substituting to prevent cadmium effluent from mining and smelting is to import not only cadmium but all other minerals and other commercially important substances derived from the ore or to replace every one of these substances in all end use products. Obviously, shutdown is the value of the industry, since all plants have effluent and all would be closed. This would entail annual loss of $16 million in direct value of technology. From entry A1, Table 50, you can see that the effectiveness of going from best available control technology to shutdown ranges from 1.3×10^{10} per health effect to infinity, with 1.3×10^{12} per health effect at the mid-range. The impact on low-flow plants is slightly more favorable (entry A2) at $110 million per health effect at maximum level

Table 51
IMPACT OF ZERO RISK EFFLUENT CONTROLS FOR ELECTROPLATING

Industry	Health effects reduced over maximum control[a]	Value of technology		Cost-effectiveness[a]	
		No substitution	Substition 90%	Marginal no substitution ($/he)[b]	Marginal substitution 90% ($/he)[b]
Job shops		3×10^9	3×10^8		
Average flow	0.03			7.1×10^{12}	7.1×10^{11}
Low flow	0.4			5.3×10^{10}	5.3×10^9
Overall	0.43			5.4×10^{11}	5.4×10^{10}
Pb manufacture		4×10^8	4×10^7		
Average flow	3×10^{-3}			1.1×10^{13}	1.1×10^{12}
Low flow	3×10^{-3}			1.1×10^{12}	1.1×10^{11}
Overall	6×10^{-3}			6.1×10^{12}	6.1×10^{11}
Captive shops		6×10^9	6×10^8		
Average flow	1			4.7×10^{10}	4.7×10^9
Low flow	30			1.6×10^8	1.6×10^7
Overall	31			1.7×10^9	1.7×10^8
Total	31.44	9.4×10^9	9.4×10^8	Incremental 3×10^8	Incremental 3×10^7

[a] Maximum health effect case; multiply last two columns by 10^2 for best estimate case.
[b] Dollars per health effect

and 1.1×10^{10} per health effect at the mid-range. Moreover, control reduces total exposure to cadmium from plants on average or larger rivers by only 9%, but removes 90% of the exposure from plants on low-flow streams (Table 31).

2. Electroplating

The only substitute providing electrochemical protection is zinc, which is bulkier than cadmium. If we assume 90% of the current cadmium used in electroplating could be replaced by zinc (without considering deficiencies), then, based upon entries B, C, and D in Table 50, we can summarize the value of technology lost and the marginal cost-effectiveness of health effect reduction. This is shown in Table 51 for the maximum health estimate. Without substitutes, direct losses approximate $9 billion/year and $900 million, if 90% replacement were possible. This does not include translocation, unemployment, retraining, and indirect expenses. Moreover, the most optimistic forecast of marginal cost-effectiveness is about $16 million per health effect. It is very much higher in all other forecasts.

3. Pigments

Alternatives for pigments containing cadmium lack stability and brilliance. Moreover, one of these, lead chromate, appears to present a higher risk. The total value of the industry is more than $200 million annually. Even if only 1% of the industry were closed, and the remainder turned to substitutes, the marginal cost-effectiveness of health effect reduction would be 1.6×10^{16} per health effect and 1.8×10^{14} per health effect for average-flow and low-flow conditions, respectively. See entries E1 and E2, Table 50, for the full cost-effectiveness values.

4. Plastics, Ni-Cd Batteries, and Iron and Steel Industries

As shown in Table 50, the value of technology is $30, $104, and 4.2×10^4 million for the plastics, Ni-Cd battery, and iron and steel industries, respectively. Even at 99% substitution, cost-effectiveness of risk reduction is very high, at least as large as the values for pigments.

Table 52
AMBIENT WATER QUALITY

Industry	Population affected ($\times 10^6$)	Replacement water cost ($)	Health effects reduced (worst case)	Incremental worst estimate ($/he)[a]	Incremental best estimate ($/he)[a]
Mining and smelting					
Average flow	2.25	1.8×10^8	0.1	2×10^{10}	2×10^{12}
Low flow	0.25	2.0×10^7	3.5	6×10^6	6×10^8
Overall	2.5	2.0×10^8	3.6	6×10^7	6×10^9
Electroplating					
Average flow	90	7.4×10^9	1	7×10^9	7×10^{11}
Low flow	10	8.2×10^8	30	3×10^7	3×10^9
Overall	100	8.2×10^9	31	3×10^8	3×10^8
Phosphate fertilizer					
Average flow	30	2.5×10^9	0.1	2×10^{10}	2×10^{12}
Low flow	3.3	2.7×10^8	3.4	8×10^7	8×10^9
Overall	33.3	2.7×10^9	3.5	8×10^8	8×10^{10}
Pigments, plastics, iron and steel, Ni-Cd batteries					
Average flow	63	5.2×10^9	2×10^{-5}	3×10^{14}	3×10^{16}
Low flow	7	5.7×10^8	6×10^{-4}	1×10^{12}	1×10^{14}
Overall	70	5.7×10^9	6×10^{-4}	1×10^{13}	1×10^{15}
Overall[b]	100	8.2×10^9	37	2×10^8	2×10^{10}

[a] Dollars per health effect
[b] Assumes all sources in the same rivers, etc.

5. Phosphate Fertilizers

There are available substitutes for phosphate fertilizers, but using phosphates with low cadmium content would also reduce effluent. A total value of technology is $3.7 billion/year. The cost-effectiveness of health effects reduction is 3.7×10^{11} per health effect and 1.8×10^9 per health effect for the average- and low-flow case, respectively, of the maximum health effect level.

B. Ambient Standards (Water Quality)

Table 52 summarizes conditions for alternate water uses, should ambient standards be exceeded. This was calculated on the same basis as for chronic risks, except that cancer rather than kidney disease was used. On a zero risk basis, alternate use is required for all uses. The incremental cost-effectiveness of using cadmium-free alternate water sources (assuming they exist), is shown for average- and low-flow conditions and for a situation where we include both conditions. The overall calculation (lower row) assumes that finding alternate water supplies for 100 million people would solve the problem of all cadmium-emitting industries at once. In none of these situations is the incremental cost-effectiveness less than $6 million per health effect assuming worst case health impact. One must conclude that ambient water quality control does not effectively reduce possible cancers, particularly since the ambient level (average 5.1 ppb) is higher than the contribution of many industries.

C. Drinking Water Standards

The case for drinking water control is shown in Table 53. It is similar to Table 52, but uses a cost of $14 per person (see chronic risks) and a reduction factor of 30. Resultant concentrations are in the right hand column. Although they have very high cost-effectiveness, these controls are more effective than those based on ambient water conditions. They do not, however, result in zero risk.

Table 53
CADMIUM DRINKING WATER CONTROL EFFECTIVENESS

Industry	Population affected (× 10⁶)	Control cost ($14/person × 10⁶)	Health effects reduced (D.F. = 30)[b]	Cost-effectiveness ($/he)[a] Worst case	Best estimate	Resulting level (ppb)
Mining and smelting						
Average flow	2.25	32	0.1	3×10^8	3×10^{10}	0.3
Low flow	0.25	4	8.4	1.2×10^6	1.2×10^8	35
Overall	2.5	36	3.5	1×10^7	1×10^9	—
Electroplating						
Average flow	90	1,300	1	1.3×10^9	1.3×10^{11}	0.15
Low flow	10	140	29	5×10^6	5×10^9	20
Overall	100	1,400	30	5×10^7	5×10^9	—
Phosphate fertilizer						
Average flow	30	400	0.1	4×10^9	4×10^{11}	0.08
Low flow	3.3	50	3.4	1.5×10^7	1.5×10^9	10
Overall	33	450	3.5	1.3×10^8	1.3×10^{10}	—
Pigments, plastics, iron and steel, Ni-Cd batteries						
Average flow	63	880	2×10^{-5}	4×10^{13}	4×10^{15}	3×10^{-4}
Low flow	7	100	6×10^{-4}	2×10^{11}	2×10^{13}	0.05
Overall	70	980	6×10^{-4}	2×10^{12}	2×10^{14}	—
Overall	100	1,400	36	4×10^7	4×10^9	2

[a] Dollars per health effect
[b] D.F. = decontamination factor, i.e., the amount of pollutant removed

D. Summary — Zero Risk (Water Pathways Only)

It is evident that use of any zero risk concept involving a nonthreshold pollutant which is not highly potent makes virtually no sense in the water pollution area. Not only is the value of technology lost very large, even with substitutes, but the cost-effectiveness of risk reduction is ludicrously high in all cases and very uncertain in terms of health risk estimates.

II. "ZERO RISK" (ALL PATHWAYS)

Even if one could avert all water pollution risk, only a portion of the total risk would be removed. The last column of Table 50 (also see Table 31) summarizes percentages of total exposure avoided by reducing water risks to zero. Note that cadmium intake from smoking cigarettes provides orders of magnitude of higher exposure than that averted in all but the three low-flow cases treated in the chronic area. Zero risk has no meaning in this context.

III. AS LOW AS CAN BE MEASURED

The concept of "as low as can be measured" may be addressed in several ways. One method is to use the cadmium detection limit at the discharge pipe or the uptake point. Another is to assume that if the total modeled health impact is much less than one adverse health effect annually over the affected population, it is insignificant and perhaps meaningless. At a field detection limit of 0.2 ppb, we need not ponder control below this downstream level. Table 54 illustrates these conditions.

On an average-flow river only three industries, the same as the sources of chronic levels of risk, need regulation (column B). On low-flow rivers, three additional industries require regulation: pigments, plastics, and electroplating-PB manufacture.

Table 54
NEED FOR REGULATION BASED UPON MEASUREMENT DETECTION LEVELS

A	B	C	D	E	F	G
			Total health impact >0.1			
	Uptake point (>0.2 ppb)		Mid-range		Maximum health estimate	
Industry	Average flow	Low flow	Average flow	Low flow	Average flow	Low flow
Mining and smelting	Y(es)	Y	N	N	N	Y
Electroplating						
Job shops	N(o)	Y	N	N	N	Y
Pb manufacture	N	N	N	N	N	N
Captive shops	Y	Y	N	Y	Y	Y
Pigments	N	N	N	N	N	N
Plastics	N	N	N	N	N	N
Ni-Cd batteries	N	Y	N	N	N	N
Iron and steel	N	Y	N	N	N	N
Phosphate fertilizers	Y	Y	N	N	Y	Y

We will use a level of 0.1 annual health effect (fatal cancer) in the total population as a threshold. Column D shows the situation on average-flow rivers for a best estimate health impact. Regulation is not required in any of the industries. In the low-flow case (column E), only one, electroplating captive shops, is above the threshold. For the more conservative worst case on an average-flow river, two industries require control (column F), and on a low-flow river, two additional industries.

Both methods effectively eliminate the need to regulate a number of industries under different flow conditions. However, regulating to these levels for the remaining industries is an expensive proposition. Costs of control are summarized in Table 55 for the measurement limit of 0.2 ppb. Captive electroplating shops would have the highest cost. Keep in mind, however, that these same expenses could control chronic noncarcinogenic risks. Cancer reduction is an additional benefit. Instead of regulating to this level, it may be used as a threshold above which we would use other means for establishing regulations. Below the level, no regulation would be required.

The use of a health effect threshold is illustrated in Table 56 which shows what a threshold of 0.1 health effect per year would cost each industry. For the mid-range or best estimate, only the captive electroplating shops need be controlled. A level of 0.3 health effect per year does not necessitate control for any of the industries as a best estimate. For a worst case (maximum) estimate, control of mining and smelting on low-flow rivers might be cost-effective. Cancer could be controlled under any scenario if controls were implemented for noncarcinogenic chronic risks. Again, if the health effect level is only a decision point as to whether or not to regulate, one might use other criteria to determine needed degree of control.

IV. AS LOW AS CAN BE CONTROLLED

The maximum control without shutdown is underlined in the left hand column of Table 50. In only the mining and smelting and phosphate fertilizer industries, both under low-flow conditions, is marginal cost-effectiveness for maximum health effect estimates less than $1 million per health effect.

Table 55

COST OF CADMIUM CONTROL DOWN TO THE DETECTION LIMIT
(0.2 PPM DOWNSTREAM)

Control case	Industry	Required reduction factor	Control costs (or shutdown cost) ($ × 10^6)	Health effects reduced		Incremental cost-effectiveness[a]	
				Best est.	Maximum	Best est.	Maximum
Average flow	Mining and smelting	90	2.6		0.1	10^9	10^7
	Electroplating — captive shops	33	4,630		1	10^{11}	10^9
	Phosphate fertilizers	30	18		0.4	10^9	10^7
	Overall		4,650	0.015	1.5	10^{11}	10^9
Low flow	Mining and smelting	30+	5	c	3.5	10^8	1.5×10^6
	Electroplating job shops	100	88	c	0.4	10^{10}	10^8
	captive shops	33+	605	c	30.0	10^9	10^7
	Ni-Cd batteries	7	0.9[b]	c	0.01[b]	10^{10}	10^8
	Iron and steel	2	12	c	0.01[b]	10^{10}	10^8
	Overall		100 610	0.34	34	10^9	10^7

[a] Rounded to nearest order of magnitude above $10^7 per health effect.
[b] $90,000 per plant and a minimum health impact of .01.
[c] Maximum est. $\times 10^{-2}$.

Maximum control is possible, of course, but is certainly not very cost-effective. It is meaningless if cadmium is not a carcinogen.

V. AS LOW AS BACKGROUND

The average background level of 5 ppb presents a risk to an individual of about 2×10^{-6} in his or her lifetime as a worst case and about 2×10^{-8} as the mid-range case.

It we assume a 10-ppb level above local background as the limit of variations, the only cases requiring control are the same as those found for noncarcinogenic chronic risks on low-flow rivers, with an average level of about 15 ppb.

Table 56
COST OF CONTROL DOWN TO A HEALTH EFFECT LIMIT (0.1 HEALTH EFFECT PER YEAR — TOTAL POPULATION)

Control case	Industry	Required reduction factor	Control costs		Health effect reduced		Incremental cost-effectiveness[a]	
			Control ($ × 10^6)	Value of technology ($ × 10^6)	Mid-range	Maximum	Mid-range ($/he)[d]	Maximum ($/he)[d]
Average flow								
	Electroplating	14	3900		c	0.9	c	10^9
	Captive shops							
	Phosphate fertilizers	30	18		c	1.4	c	10^7
	Overall		3920			2.3		10^9
Low flow								
	Mining and smelting	90	0.74		c	3.4	c	2 × 10^5
	Electroplating —							
	Job shops	33	7.4	605	c	0.4	c	10^8
	Captive shops	30 + (14)[b]	(390)[b]		0.3	30	10^9	10^7
	Phosphate fertilizers	30 +		370	c	5	c	10^8
	Overall		79(390)[b]	975	0.3	39	10^9	10^7

a   Nearest order of magnitude above 10^7.
b   Best estimate case.
c   Control not required.
d   Dollars per health effect

Chapter 28

ACCEPTABLE RISK — CADMIUM CANCER RISK

*The first precept was never to accept a thing as true
until I knew it as such without a single doubt.*

Rene Descartes
Le Discours de la Methode (1637), I

Perhaps we don't completely accept many positions or explanations of processes as true as much as we believe them to contain more truth or equity than the alternatives. In essence, that is the rationale of risk/benefit balancing, accuracy within study limitations with as much equity as possible for all concerned.

Acceptance of the end product hinges on many factors, only one of which is perceived truth. This chapter, similar to Chapter 24 for noncarcinogenic chronic cadmium risks, addresses how methods for setting standards for acceptability relate to the risk of cancer from cadmium in water.

We obtained risk to individuals from a given exposure (intake) concentration directly by the dose-effect relationships shown in Figure 16. This depends upon the estimate used. Of course, the zero effect level does not require controls under any scenario.

I. COMPARABLE RISKS

Risk can be determined whether no control or any degree of control is imposed. One may then compare risk from resultant levels to other societal risks. Table 57 summarizes the risk levels to the nearest order of magnitude for industries where the worst case health estimate for no control is above 10^{-8} health effects per lifetime.

On a worst case health effect basis, plants on average-flow rivers produce effects no worse than conditions listed in Table 45, the societal risks which increase the chance of death by 0.000001. For a mid-range case, they are below these levels by a factor of 100. The same three industries with chronic risk from low-flow sites also have higher risk. However, as a best estimate only mining and smelting cause such a problem. The exercise of the first level of control reduces the worst case health impact to that equivalent to smoking about 15 cigarettes per year (cancer, heart disease). As a best estimate, risks are all below the 10^{-6} per lifetime level.

II. ARBITRARY RISK NUMBERS

Given the acceptance of our worst case estimates, a level of 10^{-5} health effects per lifetime means that only low-flow sites for mining and smelting, captive electroplating shops, and phosphate fertilizers need be controlled and these only to the first level. No controls are needed at the mid-range estimate.

At 10^{-6} health effects per lifetime, the same three industries must be controlled for low flow as a worst case, but only mining and smelting as a mid-range estimate. Moreover, first-level controls do not reach the 10^{-6} health effects per lifetime level for mining and smelting and captive electroplating shops of low-flow rivers.

At 10^{-7} health effects per lifetime, all must be controlled as a worst case but only the three mentioned above must contemplate controls on low-flow sites as a mid-range estimate.

Table 57

CANCER RISKS FROM INDUSTRIAL CADMIUM EFFLUENTS (TO NEAREST ORDER OF MAGNITUDE ABOVE 10^{-8}/YEAR)

| | No control level | | First level control applied | |
| | Mid-range | Worst case | Mid-range | Worst case |
Industry	(he/Lifetime)[a]	(he/Lifetime)[a]	(he/Lifetime)[a]	(he/Lifetime)[a]
Mining and smelting				
Average flow	10^{-8}	10^{-6}	10^{-9}	10^{-7}
Low flow	10^{-5}	10^{-3}	10^{-7}	10^{-5}
Electroplating				
Job shops	10^{-8}	10^{-6}	10^{-9}	10^{-7}
Low flow				
Captive shops				
Average flow	10^{-8}	10^{-6}	10^{-8}	10^{-6}
Low flow	10^{-6}	10^{-4}	10^{-7}	10^{-5}
Phosphate fertilizer				
Average flow	10^{-8}	10^{-6}	10^{-10}	10^{-8}
Low flow	10^{-6}	10^{-4}	10^{-8}	10^{-6}

[a] Health effects per lifetime

III. A SET VALUE FOR DOLLARS TO BE SPENT

If a value of 1% or 10% of the total value of cadmium discharging industries is set as a control limit, one can then determine overall costs and health-effect reduction to be attained across the industry on an equitable basis. In the three major discharging industries alone, value of technology is about $1 billion/year. If 1% of this, i.e., $10 million/year, is used to control the low-flow cases, it could easily cover the mining and smelting and phosphate fertilizer industries, but not captive electroplating shops. This is true even at 10%. The figures are identical to those for chronic risk shown in Table 47 except that cost-effectiveness based upon cancer alone is very high (right hand columns of Table 50).

IV. A SET VALUE FOR RISK REDUCTION

If a threshold of 1 health effect per year is established as a worst case (0.01/year on a best estimate), it would be the same approach as that used in the measurement limit method, but at a higher value. The same three industries on the low-flow rivers could be involved at this higher level and reached by existing controls. Certainly any such number is possible, but we must make health risk/economic tradeoffs to establish the appropriate level.

Since risks are linearly reduced by controls, this approach makes little sense on a risk basis alone.

Chapter 29

ECONOMIC AND DESIGN APPROACHES — CADMIUM CANCER RISK

Annual income twenty pounds, annual expenditure
nineteen six, result happiness,
Annual income twenty pounds, annual expenditure
twenty pounds ought and six, result misery.

Charles Dickens
David Copperfield, Ch. 12 (1849)

Decisions impacting economics and decisions based solely on economic criteria may be more difficult in the former and somewhat less relevant in the latter instance for 20th Century regulators and "regulatees" than for a character in Dicken's 19th Century novel.

Likewise, use of design approaches alone has certain limitations. In this chapter, we apply both economic and design approaches to methods of setting standards to reduce cancer risk from cadmium.

I. ECONOMIC APPROACHES

As in Chapter 25 when considering chronic cadmium risks, we will address three different economic approaches: the marginal cost of risk reduction, tax incentives, and risk/cost/benefit balancing. Similarly, we discuss use of the design approaches BPT and BAT in setting standards for cadmium cancer risk.

A. Marginal Cost of Risk Reduction

On the basis of mid-range estimates, there are no conditions under which effluent, water quality, or drinking water standards are cost-effective below $10 million per health effect reduced (as seen in Tables 50, 52, 53). On a worst estimate basis, two industries fall below $1 million per health effect. Both are on low-flow sites with the first level of effluent control:

1. Mining and smelting — 1.2×10^5 per health effect
2. Phosphate fertilizer — 4.5×10^5 per health effect

In the remaining scenarios, values for cost-effectiveness range well up to 10^{16} per health effect (worst estimate). Control in water for a weak carcinogen such as cadmium, except perhaps for a few special cases, does not appear to be cost-effective. Fortunately, in some instances, controls are cost-effective in reducing noncarcinogenic chronic risks. In these cases, the marginal cost of cancer risk reduction is close to zero and comes as a bonus.

B. Tax Incentives

If one could set an effluent tax of $100,000 to $500,000 per health effect (worst case) to reduce concentration of discharge (based upon our health impact model), it could be either spread among all plants or applied to only those on low-flow rivers. If the aim is to eventually force controls, the tax should exceed control costs. Calculations in Table 58 present a break-even value in which the tax exactly equals the control cost per plant.

This tax would be preferred by the electroplating industry under all conditions. Preference in the other industries depends upon the base rate and whether the total industry or only low-flow plants are taxed. The cost-effectiveness of ambient and drinking water control make taxing for cleanup out of the question.

Table 58

EFFLUENT TAXES AS CONTROL INCENTIVE FOR CADMIUM INDUSTRIES

Industry	Control costs per plant	Health effects reduced	Tax rate per plant[a] $100 K/he[b]	$250 K/he[b]	$500 K/he[b]	Break-even
Mining and smelting (first-level control)	$190,000	3.5				
Total industry			$50 K	$125 K	$250 K	380 K/he[b]
Low-flow plants			175 K	444 K	890 K	108 K/he[b]
Electroplating	$750,000	30				
Captive shops (second-level control)						
Total industry			$0.5 K	$1.25 K	$2.5 K	1.5×10^8/he[b]
Low-flow plants			5 K	12.5 K	25 K	1.5×10^7/he[b]
Phosphate fertilizer (first-level control)	$200,000	5				
Total industry			$5.4 K	$14 K	$27 K	3.7×10^6/he[b]
Low-flow plants			54 K	140 K	270 K	370 K/he[b]

[a]   K = 1000.
[b]   1000 per health effect

Table 59

**MARGINAL COST-EFFECTIVENESS OF EFFLUENT CONTROLS
FOR LOW-FLOW RIVER SITES (CADMIUM) (WORST CASE
HEALTH ESTIMATES)**

Industry	Level of control	Slope (marginal cost-effectiveness) 100 K/he[a]	250 K/he[a]	500 K/he[a]
Mining and smelting				
Low flow	1st	1.2	0.5	0.25
	2nd	45	18	9
	Shutdown	1×10^3	4×10^2	2×10^2
Electroplating				
Captive shops	1st	1×10^2	40	20
Low flow	2nd	5×10^2	2×10^2	1×10^2
	Shutdown	1×10^3	4×10^2	2×10^2
Phosphate fertilizer				
Low flow	1st	4.5	1.8	0.9
	Shutdown	2×10^4	8×10^3	4×10^3

[a]   1000 per health effect

C. Risk/Cost/Benefit Balancing

By definition, control costs have a marginal value of technology of unity on direct costs until the industry is shut down. For indirect costs, an added factor of 2 below shutdown (a slope of 1/2 and a factor of 10 after shutdown (a slope of 0.1) is consistent with the approach we used for noncarcinogenic chronic risks. An equivalent value for health effect reduction may be found by dividing the values of marginal cost-effectiveness in the right-hand column of Tables 50, 52, and 53 by the estimated value of life, e.g., $100,000; $450,000; or $500,000. For example, the marginal cost-effectiveness for the first level of control for mining and smelting is 1.3×10^7 per health effect (worst case from Table 50). At $100,000 per health effect, the marginal cost effect is 1.3×10^2, greater than either 1, 0.5, or 0.1 and requiring no control. Table 59 shows the slope of the cost-effectiveness curve of controls for three values of health effect reduction. For a slope of 0.5, taking into account the indirect costs of investment in controls (i.e., reduced productivity), we need only the first control level after mining and smelting of $500,000 per health effect. At unity slope, first-level control for phosphate fertilizer plants would be added at $500,000 per health effect and for mining and smelting at $250,000 per health effect. One need not ponder industry shutdown since cost-effectiveness exceeds a slope of 0.1 in every case.

Assuming reasonable slopes for the marginal value of technology in direct and indirect terms, this approach seems to work for cancer effects. Contrasted with the noncarcinogenic chronic case, where the nonlinearities overwhelmed the method, it provides perspective on a more linear basis (cancer effects are assumed to be linear). However, if one attempts to use the mid-range health estimates, the process breaks down. It is entirely sensitive to the uncertainty in health impact.

II. DESIGN APPROACHES

The design approaches BPT and BAT are the same whether chronic risks are cancer or not. Only the cost-effectiveness measure differs. For cancer, all such controls are extremely cost-ineffective except perhaps first level on low-flow sites for the mining and smelting and the phosphate fertilizer industries (worst case). For the mid-range cases, controls are not cost-effective under any conditions.

Chapter 30

SUMMARY — CADMIUM CANCER RISK

When you can measure what you are speaking about,
and express it in numbers, you know something
about it; but when you cannot measure it, when
you cannot express it in numbers, your
knowledge is of a meager and unsatisfactory
kind: it may be the beginning
of knowledge, but you have scarcely,
in your thoughts, advanced to
the stage of science.

William Thomson (Lord Kelvin)
Popular Lectures and Addresses (1891—1894)

Here we are concerned not only with science, but with trans-science, that range of concerns in the social and political spheres to which we apply scientific input. Although difficult or impossible to quantify, these issues are nonetheless real and important in assessing methods of setting appropriate risk levels.

The final selection of a standard for cadmium or any other hazardous substance is a judgment based on these and other considerations. A discussion of that selection is beyond the scope of this book. We have, however, estimated certain resultant risks and costs of employing various approaches to setting levels for standards, as we have here for cancer risks of cadmium.

Cadmium is, at worst, a weak carcinogen when compared, for example, to arsenic. Nearly all level setting methods are extremely sensitive to health effect estimates. Sensitivity between the worst and the mid-range estimate is very high and, of course, all bets are off if cadmium is not a carcinogen. The industries presenting noncarcinogenic chronic risks are also those associated with the highest cancer risk. As we have said, controls to reduce chronic risks also help reduce cancer risks.

Zero risk has no meaning for a nonthreshold pollutant in the cases addressed. As long as polluting industries exist, there are no perfect water discharge controls. First, even with worst case estimates, the cost-effectiveness is, at the least, in the millions of dollars for each health effect eliminated. While we did not attempt to select a cost-effective criterion, these are unreasonably high levels. This applies to ambient and drinking water standards as well as effluent controls. Moreover, control where there is average flow reduces only a very small portion of the risk of cancer. Exposure for smoking cigarettes is orders of magnitude higher.

The as low as can be measured approach does not help in setting standards, but does provide a rationale for *de minimus* levels below which we need not regulate. Such a limit could be set on either a measurement basis or by use of health effect estimates, providing a major administrative tool to tell us when to stop contemplating controls where there is minimal impact. The actual choice is a value judgment which the relevant authorities must make. It could be justified on the basis of reduction in administrative expense vs. the value of averted health effects. The latter would be assigned on a judgment basis.

As low as can be controlled is here not only cost-ineffective, but meaningless as cadmium is either only a very weak carcinogen or not carcinogenic at all. Unlike purchasing margins of safety for chronic risks, the cost-effectiveness for the best controls can be calculated, and may be meaningful in some cases (mining and smelting, phosphate fertilizer plants on low-flow sites).

As low as background, depending on the background measure, results only in control of low-flow site industries, although it provides a means of discriminating between average- and low-flow cases and could be useful in reducing risk to low, but not zero, levels.

Risk comparisons provide perspective for judging which *de minimus* levels to set on a low as can be measured or a low as background case. One may also use arbitrary risk numbers based upon both exposure and individual risk, i.e., total health effects, but it is unlikely that individual risk levels would be very useful if exposure is ignored.

A set value for dollars to be spent only works where control expenses are but a small fraction of the value of technology. Thus, an approach of this nature must be situation-specific, resulting in diverse control strategies across and within industries. Marginal operators will be forced out of business in many instances, resulting in reduced competition and uneven application of costs in terms of health impact.

Using marginal cost of risk reduction is one way of setting standards for different industries. The choice depends upon whether and what published value is selected for avoiding a premature death.

Tax incentives based upon effluent discharges might work in some situations but, in general, rates would be very high and difficult to administer. Risk/cost/benefit balancing is effective to some extent for cancer, but only because the number of effects is small compared with the value of technology and values of life. It is not particularly useful when shutdown is possible, or probable, as it may be for the electroplating industry. Because indirect costs and benefits of shutdown produce ripple effects throughout the economy, one must take into account the imperfect nature of substitution. All of these solutions involve political decisions.

The design approaches BPT and BAT by themselves cannot be practically applied to the total industry, whether or not cadmium is a weak carcinogen. It might be practical under special conditions such as low flow or in conjunction with *de minimus* evidence, but not as a blanket policy.

Part IX

EVALUATING APPROACHES TO STANDARDS FOR PHENOL CANCER RISK

Man can learn nothing unless he proceeds from the
known to the unknown

Claude Bernard
Bulletin of New York Academy of Medicine
Vol. IV, p. 997

This saying is as appropriate to scientific investigation as it is to learning. We have carefully constructed our analysis, building upon obtained and derived information to move closer to a knowledge of applicability of various methods to different situations. We may now state our hypothetical case against phenol as a cancer agent in water; and a *weak* case it is.

The risk of cancer from exposure to phenol differs from that to cadmium in many respects. First, there is but one bit of evidence of phenol carcinogenicity in animals and that at a rather high dose level. Based upon worst case projections, phenol potency must be well below that of cadmium. Moreover, it is biodegradable; cadmium is not.

The one possibility for significantly higher cancer impacts is that humans may be exposed to chlorinated phenols, some of which may be carcinogenic, formed through reaction with chlorine used in purifying water supplies. Very little data are available on the increased adverse health impact of chlorinated phenols but, for the sake of illustrating the cost-effectiveness of health risk reduction, we assumed 100 times higher potency for these compounds. In actuality, compared with other organic molecules which form chloro-organic substances, there are only very low levels of phenol in water supplies.

In many cases with cadmium, chronic noncancer risks necessitated some form of pollution control. Cancer risks are reduced as a dividend. Because there are no noncarcinogenic chronic phenol risks from water, all controls must be set on the basis of either human cancer risk or protection of aquatic species. However, cleanup of water supplies to extract other chlorinated hydrocarbons also reduces phenols since removal methods are the same.

The existence of a detectable odor limit provides a unique reference point for phenol in water. It is both a low-cost and directly observable test point at a reasonably low level.

Table 60 summarizes the results of applying effluent controls on each industry for the three categories of river flow. Reference will be made to this table throughout the chapters comprising Part IX.

Table 60
EFFLUENT CONTROL APPLIED TO PHENOL FOR CANCER IMPACT

Industry (Number of plants)	Pipe level concentration	Downstream concentration	Reduction factor	Incremental ($ million)	Incremental value of technology ($ million)	Health effects reduced[a] Maximum	Mid-range	Minimum	Marginal cost of risk reduction Maximum ($/he)[a]	Mid-range ($/he)[a]	Minimum ($/he)[a]
Coke ovens (no dephenolization)											
A1 Average-flow (38)	3,360 ppm	85 ppb	0	0	(527)	(0.2)	$\times 10^{-2}$[c] (0.2)	0			
level 1	47	1.2	72	19[b]		0.19722	0.19722	0	9.6×10^{7}	9.6×10^{9}	9.6×10^{9}
2	9	0.2	360	8		0.00222	0.00222	0	3.6×10^{9}	3.6×10^{11}	3.6×10^{11}
3	0.1	0.002	36,000	28	434	0.00055	0.00055	0	5.1×10^{10}	5.1×10^{12}	5.1×10^{12}
4	0	0				0.0001	0.00001	0	4.3×10^{13}	4.3×10^{15}	4.3×10^{15}
A2 Low-flow (4)	3,360 ppm	11,600 ppb	0	0	(53)	(2)	(0.02)	0			
level 1		161	72	2[b]		1.0722	0.019722	0	9.6×10^{5}	9.6×10^{7}	9.6×10^{7}
2		32	360	1		0.0222	0.000222	0	3.6×10^{7}	3.6×10^{9}	3.6×10^{9}
3		0.3	36,000	3	43	0.0055	0.000055	0	5.1×10^{8}	5.1×10^{10}	5.1×10^{10}
4		0				0.0001	0.000001	0	4.3×10^{11}	4.3×10^{13}	4.3×10^{13}
Coke ovens (with dephenolization)						$\times 10^{-3}$[c]	$\times 10^{-5}$[c]				
B1 Average-flow (25)	71 ppm	2.0 ppb	0	0	(310)	(5)	(5)	0	—		
level 1	14	0.4	5	5		4.00	4.00	0	1.3×10^{9}	1.3×10^{11}	1.3×10^{11}
2	0.7	0.02	100	18	287	0.95	0.95	0	1.9×10^{10}	1.9×10^{12}	1.9×10^{12}
3	0	0				0.05	0.05	0	5.7×10^{12}	5.7×10^{14}	5.7×10^{12}
B2 Low-flow (3)	71 ppm	240 ppb	0	0	(31)	$\times 10^{-2}$ (3)	$\times 10^{-4}$ (3)	0	—		
level 1		48	5	0.5		2.40		0	2.1×10^{7}	2.1×10^{11}	2.1×10^{11}
2		2	100	2.0	29	0.57		0	3.5×10^{8}	3.5×10^{10}	3.5×10^{10}
3		0				0.03		0	9.7×10^{10}	9.7×10^{12}	9.7×10^{12}
Phenol manufacturing											
C1 Average-flow (14)	960 ppm	94 ppb	0	0	(934)	$\times 10^{-2}$ (7)	$\times 10^{-4}$ (7)	0			
level 1	19	2	50	40		6.860	6.860	0	5.8×10^{8}	5.8×10^{8}	5.8×10^{10}
2	0.2	0.02	5,100	80	814	0.139	0.139	0	5.8×10^{10}	5.8×10^{12}	5.8×10^{12}
3	0	0				0.001	0.001	0	8.1×10^{13}	8.1×10^{15}	8.1×10^{15}
C2 Low-flow (2)	960 ppm	12,600 ppb	0	0	(133)	(10.0)	(0.1)	0			
level 1		250	50	6		9.800	0.0980	0	6.1×10^{5}	6.1×10^{7}	6.1×10^{7}
2		2	5,100	11	116	0.198	0.00198	0	5.6×10^{7}	5.6×10^{9}	5.6×10^{9}
3		0				0.002	0.00002	0	5.8×10^{10}	5.8×10^{12}	5.8×10^{12}

Source / level	(ppm)	(ppb)								
Petrochemical										
D1 Average-flow (453)	367 ppm	383 ppb		(11,000)	(3)	(0.03)				
level 1	7.3	8	0	0		2.940	0.02940	0	4.4×10^{8}	4.4×10^{10}
2	0.07	0.08	50	1,300	7,100	0.059	0.00059	0	4.4×10^{10}	4.4×10^{12}
3	0	0	5,100	2,600		0.001	0.00001	0	7.1×10^{12}	7.1×10^{14}
D2 Low-flow (45)	367 ppm	45,300 ppb		(1,100)	(140)	(1.4)				
level 1		900	0	0		137.20	1.3720	0	9.5×10^{5}	9.5×10^{7}
2		9	50	130	710	2.77	0.0277	0	9.4×10^{7}	9.4×10^{9}
3		0	5,100	260		0.03	0.0003	0	2.4×10^{10}	2.4×10^{12}
Oil refineries										
E1 Average-flow (349)	57 ppm	35 ppb		(92,000)	(0.3)	$\times 10^{-2}$ (0.3)				
level 1	1	0.7	0	0		0.29400	0.29400	0	3.4×10^{9}	3.4×10^{11}
2	0.01	0.007	50	1,000	89,000	0.00594	0.00594	0	3.4×10^{11}	3.4×20^{13}
3	0	0	5,100	2,000		0.00006	0.00006	0	1.5×10^{15}	1.5×10^{17}
E2 Low-flow (35)	57 ppm	4,400 ppb		(9,200)	(14)	$\times 10^{-2}$ (0.14)				
level 1		88	0	0		13.720	0.13720	0	7.3×10^{6}	7.3×10^{8}
2		0.9	50	100	8,900	0.277	0.00277	0	7.2×10^{8}	7.2×10^{10}
3		0	5,100	200		0.003	0.00003	0	3.0×10^{12}	3.0×10^{14}
Plywoods										
F1 Average-flow (477)	2.3 ppm	1.0×10^{-3} ppb		(5,200)	$\times 10^{-5}$ (1)	$\times 10^{-7}$ (1)				
level 1	0.05	2.0×10^{-5}	0	0		0.9800	0.9800	0	1.4×10^{14}	1.4×10^{16}
2	0.0005	2.0×10^{-7}	50	1,400	1,100	0.0198	0.0198		1.4×10^{16}	1.4×10^{18}
3	0	0	5,100	2,700		0.0002	0.0002		5.5×10^{17}	5.5×10^{19}
F2 Low-flow (48)	2.3 ppm	1 ppb		(520)	$\times 10^{-3}$ (1)	$\times 10^{-5}$ (1)				
level 1		0.02	0	0		0.9800	0.9800	0	1.4×10^{11}	1.4×10^{13}
2		0.0002	50	140	110	0.0198	0.0198		1.4×10^{13}	1.4×10^{15}
3		0	5,100	270		0.0002	0.0002		5.5×10^{14}	5.5×10^{16}
Blast furnaces										
G1 Average-flow (251)	0.34 ppm	0.72 ppb		(24,000)	$\times 10^{-4}$ (9)	$\times 10^{-6}$ (9)				
level 1	0.007	0.01	0	0		8.820	8.820	0	7.9×10^{11}	7.9×10^{13}
2	0.00007	0.0001	50	700	21,900	0.178	0.178	0	7.9×10^{13}	7.9×10^{15}
3	0	0	5,100	1,400		0.002	0.002	0	1.1×10^{17}	1.1×10^{19}
G2 Low-flow (25)	0.34 ppm	75 ppb		(2,400)	$\times 10^{-2}$ (0.1)	$\times 10^{-2}$ (0.1)				
level 1		1.5	0	0		0.09800	0.09800	0	7.1×10^{8}	7.1×10^{10}
2		0.015	50	70	2,190	0.00198	0.00198	0	7.1×10^{10}	7.1×10^{12}
3		0	5,100	140		0.00002	0.00002	0	1.1×10^{14}	1.1×10^{16}

Table 60 (continued)
EFFLUENT CONTROL APPLIED TO PHENOL FOR CANCER IMPACT

Industry (Number of plants)	Pipe level concentration	Downstream concentration	Reduction factor	Incremental ($ million)	Incremental value of technology ($ million)	Health effects reduced[a] Maximum	Health effects reduced[a] Mid-range	Health effects reduced[a] Minimum	Marginal cost of risk reduction Maximum ($/he)[a]	Marginal cost of risk reduction Mid-range ($/he)[a]	Marginal cost of risk reduction Minimum ($/he)[a]
Phenolic resins							$\times 10^{-2}$				
H1 Average-flow (41) level 1	1600 ppm	154 ppb	0	0	(570)	(0.4)	(0.4)	0	3.1×10^{8}	3.1×10^{10}	
$\frac{2}{3}$	32	3	50	120		0.3920	0.3920	0	3.0×10^{10}	3.0×10^{12}	
	0.3	0.03	5,100	240	210	0.0079	0.0079	0	2.1×10^{12}	2.1×10^{14}	
	0	0				0.0001	0.0001	0			
H2 Low-flow (4) level 1	1600 ppm	20,800 ppb	0	0	(57)	(21)	0.21	0	5.8×10^{5}	5.8×10^{7}	
$\frac{2}{3}$		400	50	12		20.580	0.20580	0	5.7×10^{7}	5.7×10^{9}	
		4	5,100	24	21	0.416	0.00416	0	5.3×10^{9}	5.3×10^{11}	
		0				0.004	0.00004	0			

Note: $/he = Dollars per health effect.

[a] Carried to three significant figures only to show effects. This precision should not be assumed to imply that information is meaningful.

[b] Assumes economic recovery through dephenolization will cover 67% of cost.

[c] Indicates order of magnitude for the following category.

Chapter 31

RISK AVERSION METHODS — PHENOL CANCER RISK

Confess yourself to heaven;
Repent what's past; avoid what is to come.

William Shakespeare
Hamlet, Act III

Avoidance of exposure to toxic materials in substances impossible, or virtually impossible, to shun, as drinking water, may require societal action. This action is often in the form of standards. Approaches to standards based upon an avoidance of cancer risk from phenol in water are the subject of this chapter.

I. ZERO RISK (WATER PATHWAYS ONLY)

Zero risk has been defined. At this point, we will consider how that concept, as applied to phenol in water, may be implemented in risk aversion methods of standard setting. Much of the discussion on effluent control is drawn from a summary of results in Table 60 presented in the introduction to Part IX.

A. Effluent Standards

1. Coke Plants (No Dephenolization)

In coking operations, phenol is a by-product with no economic significance unless it is recovered. We have divided the coke industry into two categories because only 41% of the existing plants utilize phenol recovery, a process known as dephenolization. Coke is a fuel used primarily in blast furnaces of the iron and steel industry as an integral part of the ironmaking operations. There is no short-term substitute. Closing these plants might seriously affect not only the iron and steel industry, but other industries with metallurgical uses of coke as well. For our purposes, we will represent cost as the production value of the coke-making industry, about $879 million/year (from Table 22). For plants not utilizing dephenolization, the value of technology is $527 million/year.

From entry A1, Table 60, one can see that the effectiveness of going from best available control technology to plant closure ranges upward from 4.3×10^{13} per health effect to infinity, with a mid-range of 4.3×10^{15} per health effect. As with many of the cadmium cases, the impact on low-flow plants is several orders of magnitude more favorable (entry A2). The mid-range marginal cost of risk reduction is 4.3×10^{13} per health effect for shutdown.

2. Coke Plants (With Dephenolization)

This category includes plants already designed to facilitate the recovery of volatile by-products from coking as opposed to those utilizing an older process (the beehive process). The loss due to shutdown is $310 million/year with mid-range marginal cost-effectiveness of 5.7×10^{14} per health effect for plants on average-flow rivers and 9.7×10^{12} per health effect for those on low-flow rivers. Thus, plants on average-flow rivers not utilizing recovery have a marginal cost only eight times that of those doing so. The ratio is 4:1 for plants on low-flow sites.

3. Phenol Manufacturing

We estimate that approximately 50% of the phenol presently used could be economically

Table 61

IMPACT OF ZERO RISK EFFLUENT CONTROLS FOR PHENOL-EMITTING INDUSTRIES

Industry (% substitutable)	Health effects reduced over maximum control	Value of technology		Cost-effectiveness[a]	
		No substitution	Substitution	Marginal no substitution ($/he)[b]	Marginal substitution ($/he)[b]
Phenol manufacturing (50%)		9.3×10^8	4.6×10^8		
Average flow	0.1			8.1×10^{13}	4.0×10^{13}
Low flow	10.0			5.8×10^{10}	2.9×10^{10}
Overall	10.1			3.1×10^{11}	1.6×10^{11}
Plywoods (40%)		5.2×10^9	2.1×10^9		
Average flow	1×10^{-5}			5.5×10^{17}	2.2×10^{17}
Low flow	1×10^{-3}			5.5×10^{14}	2.2×10^{14}
Overall	1×10^{-3}			6.0×10^{15}	2.4×10^{15}
Phenolic resins (35%)		5.7×10^8	2.0×10^8		
Average flow	0.4			2.1×10^{12}	7.4×10^{11}
Low flow	21.0			5.3×10^9	1.9×10^9
Overall	21.4			5.6×10^{10}	2.0×10^{10}

[a] Maximum health effect case; multiply last 2 columns by 10^2 for best estimate case.
[b] Dollars per health effect

replaced, but this would not significantly reduce the loss from industry shutdown. With substitution, the value of technology lost would be 4.0×10^{13} for plants on average-flow rivers, as depicted in Table 61. The marginal cost-effectiveness in the most optimistic situation (low flow with substitution) is still about $29 billion per health effect.

4. Petrochemical Industry and Oil Refineries

Because there are no short-term substitutes for petrochemical products or oil production, the economic impact of industry shutdown cannot be mitigated. Entries D and E, Table 61, show marginal cost-effectiveness of at least $24 billion per health effect for the petrochemical industry and 3×10^{12} per health effect for oil refineries.

5. Plywoods

Phenolic resin is used as a glue in the plywood industry. There is no direct substitute which does not result in a substantial loss in quality and heat resistance. However, an "indirect" substitute, lignin, utilized as an extender mixed with resin, produces an adequate product and could reduce phenolic resin use by about 40%. Plywood plants would then emit phenol in smaller volume and concentration. We assume this percent substitution would have the same effect in the phenol manufacturing industry.

Table 61 shows that, with this 40% substitution, cost-effectiveness is still 2.2×10^{17} per health effect and 2.2×10^{14} per health effect for the average-flow and low-flow cases, respectively. The marginal effect is severe because plywood plants already emit phenol in extremely low concentrations (as seen in Table 13). The reduction cost is therefore very high and the number of health effects avoided very low.

6. Blast Furnaces

Similar to the plywood industry, blast furnaces typically have emissions of low phenol concentration. There are no short-term replacements for most uses of iron and steel. Entry G in Table 60 shows marginal cost-effectiveness of 1.1×10^{17} per health effect and $1.1

Table 62
AMBIENT WATER QUALITY (PHENOL)

Industry	Population affected ($\times 10^6$)	Replacement water costs ($)	Health effects reduced (worst case)	Incremental worst estimate	Incremental best estimate
Coke ovens (no dephenolization)					
Average flow	12.6	1.0×10^9	0.2	5×10^9	5×10^{11}
Low flow	1.4	1.1×10^8	2.0	6×10^7	6×10^9
Overall	14	1.1×10^9	2.2	5×10^8	5×10^{10}
Coke ovens (with dephenolization)					
Average flow	8.1	6.6×10^8	0.005	1×10^{11}	1×10^{13}
Low flow	0.9	7.4×10^7	0.030	2×10^9	2×10^{11}
Overall	9.0	7.4×10^8	0.035	2×10^{10}	2×10^{12}
Phenol manufacturing					
Average flow	4.5	3.7×10^8	0.1	4×10^9	4×10^{11}
Low flow	0.5	4.1×10^7	10	4×10^6	4×10^8
Overall	5.0	4.1×10^8	10.1	4×10^7	4×10^9
Petrochemical					
Average flow	90	7.4×10^9	0.3	2×10^9	2×10^{11}
Low flow	10	8.2×10^8	14.0	6×10^6	6×10^8
Overall	100	8.2×10^9	14.3	6×10^7	6×10^9
Oil refineries					
Average flow	90	7.4×10^9	0.3	2×10^{10}	2×10^{12}
Low flow	10	8.2×10^8	14.0	6×10^7	6×10^9
Overall	100	8.2×10^9	14.3	6×10^8	6×10^{10}
Plywoods					
Average flow	90	7.4×10^9	1×10^{-5}	7×10^{14}	7×10^{16}
Low flow	10	8.2×10^8	1×10^{-3}	8×10^{11}	8×10^{13}
Overall	100	8.2×10^9	1×10^{-3}	8×10^{12}	8×10^{14}
Blast furnaces					
Average flow	81	6.6×10^9	9×10^{-4}	7×10^{12}	7×10^{14}
Low flow	9	7.4×10^8	0.1	7×10^9	7×10^{11}
Overall	90	7.4×10^9	0.1	7×10^{10}	7×10^{12}
Phenolic resins					
Average flow	13.5	1.1×10^9	0.4	3×10^9	3×10^{11}
Low flow	1.5	1.2×10^8	21	6×10^6	6×10^8
Overall	15	1.2×10^9	21.4	6×10^7	6×10^9
Overall	100	8.2×10^9	190	4×10^7	4×10^9

$\times 10^{14}$ per health effect for the average- and low-flow cases, respectively, at the maximum health effect level. The total value of technology is $24 billion/year.

7. Phenolic Resins

We estimate that alternative products can economically replace 35% of the present uses of phenolic resins. As with phenol manufacturing and plywood industries (Table 61), the cost impact is reduced insignificantly. The value of technology is $200 million and the marginal cost-effectiveness is $740 billion per health effect and $1.9 billion per health effect for the average- and low-flow cases, respectively, with 35% substitution.

B. Ambient Standards (Water Quality)

Table 62 shows the cost of alternate water if ambient standards are exceeded. Of course, the zero risk criterion requires alternate uses in each situation. The same calculations for cadmium indicated that $6 million per health effect was the lowest incremental cost-effectiveness, assuming worst case health impact (see Table 52). For phenol, the lowest incremental cost-effectiveness is $4 million per health effect (worst case health impact).

Table 63
PHENOL DRINKING WATER CONTROL EFFECTIVENESS

Industry	Population affected ($\times 10^6$)	Control cost ($\times 10^6$)	Health effects reduced (DF = 100)	Cost-effectiveness Worst case	Cost-effectiveness Best case	Resulting level (ppb)
Coke ovens (no dephenolization)						
Average flow	12.6	1000	0.2	5×10^9	5×10^{11}	0.8
Low flow	1.4	100	2.0	5×10^7	5×10^9	120
Overall	14	1100	2.2	5×10^8	5×10^{11}	—
Coke ovens (with dephenolization)						
Average flow	8.1	650	0.005	1×10^{11}	1×10^{13}	0.02
Low flow	0.9	70	0.030	2×10^9	2×10^{11}	2
Overall	9.0	720	0.035	2×10^{10}	2×10^{11}	—
Phenol manufacturing						
Average flow	4.5	360	0.1	4×10^9	4×10^{11}	0.9
Low flow	0.5	40	9.9	4×10^6	4×10^8	130
Overall	5.0	400	10.0	4×10^7	4×10^9	—
Petrochemical						
Average flow	90	7200	3	2×10^9	2×10^{11}	4.0
Low flow	10	800	139	6×10^6	6×10^8	450
Overall	100	8000	142	6×10^7	6×10^9	—
Oil refineries						
Average flow	90	7200	0.3	2×10^{10}	2×10^{12}	0.4
Low flow	10	800	13.9	6×10^7	6×10^9	40
Overall	100	8000	14.2	6×10^8	6×10^{10}	—
Plywoods						
Average flow	90	7400	1×10^{-5}	7×10^{14}	7×10^{16}	1×10^{-5}
Low flow	10	820	1×10^{-3}	8×10^{11}	8×10^{13}	1×10^{-2}
Overall	100	8200	1×10^{-3}	8×10^{12}	8×10^{14}	—
Blast furnaces						
Average flow	81	6600	9×10^{-4}	7×10^{12}	7×10^{14}	0.007
Low flow	9	740	0.1	7×10^9	7×10^{11}	0.8
Overall	90	7400	0.1	7×10^{10}	7×10^{12}	—
Phenolic resins						
Average flow	13.5	1100	0.4	3×10^9	3×10^{11}	0.15
Low flow	1.5	120	20.8	6×10^6	6×10^8	200
Overall	15	1200	21.2	6×10^6	6×10^9	—
Overall	100	8200	190	4×10^7	4×10^9	

For phenol, the ambient level (average 11 ppb) is much lower than industrial effluent. Based on our cost-effectiveness calculations, it appears that phenol ambient water quality control under zero risk is not a particularly effective way to reduce cancer risk.

C. Drinking Water Standards

For phenol, the technology to purify drinking water is activated carbon with a typical reduction factor of 100. Based on the process shown in Table 27, about 36,000 people could be served by such a plant. Costs are about $160 per person on a 1980 crude annual cost basis.

Using the same economies of scale assumption as for cadmium results in a cost of $80 per person. This is the same as the replacement cost used in Table 62.

The cost-effectiveness calculations in Table 63 use this $80 figure and a reduction factor of 100. Results are equivalent to those in Table 62 for water replacement, because the activated carbon method decreases adverse health effects by 99%. In terms of cost-effec-

Table 64
**NEED FOR REGULATION BASED UPON MEASUREMENT DETECTION
LEVELS (PHENOL)**

A	B	C	D	E	F	G
			Total health impact $\geqslant 0.1$			
	Uptake point >0.1 ppb		Mid-range		Maximum health estimate	
Industry	Average flow	Low flow	Average flow	Low flow	Average flow	Low flow
Coke ovens						
no dephenolization	Y(es)	Y	N	N	Y	Y
with dephenolization	Y	Y	N	N	N	N
Phenol manufacturing	Y	Y	N	N	N	Y
Petrochemical	Y	Y	Y	Y	Y	Y
Oil refineries	Y	Y	N	Y	Y	Y
Plywoods	N(o)	Y	N	N	N	N
Blast furnaces	Y	Y	N	N	N	N
Phenolic resins	Y	Y	N	Y	Y	Y

tiveness, this is essentially the same as the 100% decrease for water replacement. We conclude that drinking water control is not an effective means of abating phenol cancer risk.

Even if it is assumed that phenols are converted to chlorinated phenols with 100 times our worst case health effect estimate, the result would be orders of magnitude above $1 million per health effect for all industries on average-flow rivers. In low-flow situations, four out of eight industries in Table 63 are below this level. These values are found by subtracting two orders of magnitude from the figures in the worst case column for cost-effectiveness. On an overall basis (last row of the table), the cost is $400,000 per health effect. If chlorinated phenols are more potent, one can make further appropriate reductions.

II. ZERO RISK (ALL PATHWAYS)

For cadmium, reduction in water risk from industry wastewater contributes only a fractional reduction to total risk. With phenol, the industry wastewater sources are the predominant source of exposure to the general population. Therefore, shutdown in a phenol-emitting industry effectively reduced total risk. Including ambient conditions would not contribute to the analysis. However, if one weighs the total intake for special populations, such as medicinal users, risk can substantially exceed that from wastewater alone. For medicinal users then, water pollution represents only a small proportion of total risk.

III. AS LOW AS CAN BE MEASURED

We will again use this concept in two different approaches: using the detection limit of phenol in water (0.1 ppb) at the uptake point; and using less than 0.1 health effects as a threshold for measuring health impact. Both these approaches are used in Table 64.

Columns B and C indicate the detection limit approach for both flow conditions. At low flow, all industries must be controlled at detection limit, at average flow, only the plywood industry achieves a downstream level below detection limit.

On the basis of a health impact measurement threshold (columns D through G), fewer industries need be controlled. In the low-flow case (column F), only three industries, pe-trochemical, oil refineries, and phenolic resins are above threshold. However, the more conservative estimate (column G) places five industries above threshold. For average-flow sites (column D) only the petrochemical industry is above threshold, while the more conservative estimate (column F) places four industries above. Using dephenolization in the coke industry results in a health impact below threshold in all cases. On the other hand, the effects from petrochemical industry are always above threshold.

Because the detection limit is so low, use of the as low as can be measured concept, based solely upon use of the limit, could give an extreme (and unnecessary) level of control. The measurement concentration, however, is not a limiting condition. Several industries are below the health effect threshold (0.1 health effects), even though they create detectable levels downstream.

Expenditures for controlling down to the health effect threshold are shown in Table 65. For the mid-range or best estimate, only the petrochemical and oil refining industries require control, although the latter is marginal with respect to the health effect threshold. In terms of cost-effectiveness, three industries are within $10 million per health effect for low-flow rivers as a worst case (maximum) estimate.

IV. AS LOW AS CAN BE CONTROLLED

The maximum control without shutdown is underlined in the levels of control in the lefthand column of Table 60. All marginal cost-effectiveness for industries in average- or low-flow for the maximum or mid-range cases are above $1 million per health effect.

V. AS LOW AS BACKGROUND

The mean level for phenol in raw water, as reported by STORET, is 11.0 ppb with a standard deviation of 7.7 ppb. Thus, a background level of about 20 ppb could be set as a standard. All industries on low-flow rivers are above this level, as are those on average-flow rivers, with the exception of coke ovens (with dephenolization), plywood plants, and blast furnaces. Those exceeding 20 ppb under average-flow conditions, coke ovens with no dephenolization, phenol manufacturing, petrochemical, oil refineries, and phenolic resins, require one level of control to meet as low as background, with an incremental cost-effectiveness range between 1.0×10^8 per health effect to 3.4×10^9 per health effect for maximum health effects (see first level in Table 60).

Table 65

COST OF CONTROL DOWN TO A HEALTH EFFECT LIMIT (0.1 HE/YEAR)

Control costs	Industry	Required reduction factor	Control costs		Health effect reduced		Incremental cost-effectiveness	
			Control ($ × 10^6)	Value of technology ($ × 10^6)	Mid-range	Maximum	Mid-range ($/he)	Maximum ($/he)
Average flow	Petrochemical	50	1300		b	2.9	b	10^8
	Oil refineries	50	1000		b	0.3	b	10^9
	Phenol resins	50	120		b	0.4	b	10^7
	Overall		2420			3.6		10^8
Low flow	Phenol manufacturers	5100	17		b	10.0	b	10^6
	Petrochemical	5100(50)[a]	390(130)[a]		1.4	140.0	10^7	10^6
	Oil refineries	5100(50)[a]	300(100)[a]		0.1	14.0	10^9	10^7
	Phenolic resins	5100	36		b	21.0	b	10^6
	Overall		743(230)[a]		1.5	185.0	10^8	10^6

Note: he = health effects

a   Best estimate case.
b   Control not required.

Chapter 32

ACCEPTABLE RISK — PHENOL CANCER RISK

Always fall in with what you're asked to accept.
Take what is given, and make it over your way.
My aim in life has always been to hold my own
with whatever's going. Not against: with.

Comment
Robert Frost (1874—1963)

It is evident that the above opinion is not shared by all the citizenry. Most in this nation comply with regulations even though they may be unacceptable to them. They are, however, often able by due process to reverse these regulations. Standards are no exception. Given this, it seems prudent to achieve as much all-around acceptability for a standard as possible within the framework of our assessment.

Here we will add the application of methods to achieve acceptable standards for phenol cancer risk to that for risks from cadmium discussed in Chapters 24 and 28. As stated, we are assuming for this exercise, on the basis of very slim evidence, that phenol is carcinogenic.

I. COMPARABLE RISKS

Table 66 summarizes the risk levels (to the nearest order of magnitude) for industries with worst case health estimates above 10^{-8} health effects per lifetime if controls are not applied. Results are from the dose-effect relationship shown in Figure 23. Only plywood plants on average-flow rivers are below 10^{-8} health effects per lifetime for no control. Compared with risks from similar chemicals in society, these risks are on the low side. Based on the mid-range estimates, with the single exception of the petrochemical industry, plants on average-flow rivers are no worse than the conditions, as listed in Table 45. However, the worst case estimates are quite large and range from 10^{-1} to 10^{-6} health effects per lifetime.

After applying the first level of control, risk impact is reduced, although still large, in most industries. The low risk presented by the plywood industry and blast furnaces on average-flow rivers does not demand even first-level control. Using the second level of control, all industries on average-flow rivers are within 10^{-6} health effects per lifetime, and many do not need this level of control. Note that applying the second level is equivalent to the as low as can be controlled situation for these industries because applying the third level causes shutdown.

II. ARBITRARY RISK NUMBERS

For a level of 10^{-5} health effects per lifetime (worst case), the petrochemical and phenolic resin industries on low-flow sites require control to the second level. Likewise, at the 10^{-6} health effects per lifetime worst case estimate for coke ovens, phenol manufacturing, and oil refineries on low-flow sites, control must be to the second level, and for blast furnaces on low-flow sites, the first level. At 10^{-7} health effects per lifetime worst case, all industries except plywoods must be controlled by at least the first level and at most by the second level.

III. A SET VALUE OF DOLLARS TO BE SPENT

For all industries, the total value of technology was about $134 billion/year in 1972 ($200

Table 66
CANCER RISKS FROM INDUSTRIAL PHENOL EFFLUENTS (TO NEAREST ORDER OF MAGNITUDE ABOVE 10^{-8}/YEAR)

Industry	No control level		First-level control applied		Second level of control applied	
	Mid-range (he/lifetime)[a]	Worst case (he/lifetime)[a]	Mid-range (he/lifetime)[a]	Worst case (he/lifetime)[a]	Mid-range (he/lifetime)[a]	Worst case (he/lifetime)[a]
Coke ovens (no dephenolization)						
Average flow	10^{-6}	10^{-4}	10^{-8}	10^{-6}	—	—
Low flow	10^{-4}	10^{-2}	10^{-6}	10^{-4}	10^{-8}	10^{-6}
Coke ovens (with dephenolization)						
Average flow	10^{-7}	10^{-5}	10^{-8}	10^{-6}	—	—
Low flow	10^{-5}	10^{-3}	10^{-6}	10^{-4}	10^{-8}	10^{-6}
Phenol manufacturing						
Average flow	10^{-6}	10^{-4}	10^{-7}	10^{-5}	—	10^{-7}
Low flow	10^{-4}	10^{-2}	10^{-5}	10^{-3}	10^{-8}	10^{-6}
Petrochemical						
Average flow	10^{-5}	10^{-3}	10^{-7}	10^{-5}	—	10^{-7}
Low flow	10^{-3}	10^{-1}	10^{-5}	10^{-3}	10^{-7}	10^{-5}
Oil refineries						
Average flow	10^{-6}	10^{-4}	10^{-8}	10^{-6}	—	—
Low flow	10^{-4}	10^{-2}	10^{-6}	10^{-4}	10^{-8}	10^{-6}
Plywoods						
Low flow	10^{-8}	10^{-6}	—	—	—	—
Blast furnaces						
Average flow	10^{-8}	10^{-6}	—	—	—	—
Low flow	10^{-6}	10^{-4}	10^{-8}	10^{-6}	—	—
Phenolic resins						
Average flow	10^{-6}	10^{-4}	10^{-7}	10^{-5}	—	10^{-7}
Low flow	10^{-3}	10^{-1}	10^{-5}	10^{-3}	10^{-7}	10^{-5}

[a] Health effects per lifetime

billion/year in present figures). About 94% of this is from three industries, petrochemical, blast furnaces, and oil refineries. A 2%, but not a 1%, set value ($2 billion/year) would cover control costs in all three of these industries (from Table 60, average-flow cases). Cost-effectiveness for this method, however, is still poor.

IV. A SET VALUE FOR RISK REDUCTION

All industries using available controls meet the worst case threshold of 1 health effect per year. In fact, all could comply with a threshold of 0.3 yearly health effects which is the limiting situation for petrochemical plants on low-flow sites. Of course, one must consider cost-effectiveness from the viewpoint of a health vs. economic trade-off to determine the appropriate threshold for such a method.

Chapter 33

ECONOMIC AND DESIGN APPROACHES — PHENOL CANCER RISK

We have always known that heedless self-interest
was bad morals; we know now that it is bad economics.

Franklin Delano Roosevelt
Second Inaugural Address
January 20, 1937

Few would disagree that a degree of self-interest, of and by itself, is not wrong but that to achieve maximum long-lasting benefits, it must be balanced with concern for others or other groups. This applies to industry and government (which can lose sight of its original role), just as it does to other groups and to individuals. One must keep sight of this fact when using economic and design approaches to standards for cancer effects from phenol (as we are in this chapter) or any other type of approach to setting standards for any other hazardous substance.

I. ECONOMIC APPROACHES

As in previous chapters dealing with the application of economic approaches to other risk sources, we will here discuss each type of approach in turn.

A. Marginal Cost of Risk Reduction

As with cadmium, on the basis of mid-range estimates, there are no conditions for which effluent, water quality, or drinking water standards can be set for phenol below $10 million per health effect reduced (Tables 60, 62, 63). On a worst case basis, four situations (all on low-flow sites) fall below $1 million per health effect for the first.level of effluent control:

1. Coke ovens (no dephenolization) 9.6×10^5 per health effect
2. Phenol manufacturing 6.1×10^5 per health effect
3. Petrochemical 9.5×10^5 per health effect
4. Phenolic resins 5.8×10^5 per health effect

For the others, cost-effectiveness values range all the way to 10^{17} per health effect as a worst case estimate. These strongly parallel results for cadmium. There are only a few scenarios that might be viewed as cost-effective in controlling cancer from phenol-contaminated water.

B. Tax Incentives

We have derived some possible tax rates based upon three different health impact values. These are shown in Table 67. By comparing the break-even points, the tax equivalent of control cost to the tax rates, it is clear that these industries would prefer tax to control. The rates must be enormous to persuade them to make the economic trade-off for control rather than bearing the tax burden. In this situation, tax incentives to reduce pollution appear impractical. What would result is a charge for pollution.

C. Risk/Cost/Benefit Balancing

Table 68 shows the slope of the cost-effectiveness curves for three values of health effect reduction. These are the same values used for the cadmium cancer analysis. We again use

Table 67

**EFFLUENT TAXES AS CONTROL INCENTIVE FOR PHENOL INDUSTRIES
(WORST CASE HEALTH ESTIMATES)**

Industry	Control costs per plant	Health effects reduced	Per plant tax rate ($/plant)			
			$100 k/he	$250 k/he	$500 k/he	Break-even
Coke ovens (no dephenolization) (first-level control)	$500,000	2				
Total industry			5k	13k	26k	10^7/he
Low-flow plants			50k	130k	260k	10^6/he
Petrochemical (first-level control)	3×10^6	140				
Total industry			31k	77k	150k	10^7/he
Low-flow plants			310k	770k	1500k	10^6/he
Oil refineries (first-level control)	3×10^6	14				
Total industry			4k	10k	20k	7×10^7/he
Low-flow plants			40k	100k	200k	7×10^6/he

Note: k = 1000, he = health effects.

Table 68

**MARGINAL COST-EFFECTIVENESS OF EFFLUENT CONTROLS
FOR PHENOL ON LOW-FLOW RIVER SITES (WORST CASE
HEALTH EFFECTS)**

Industry	Level of control	Slope marginal cost-effectiveness[a]		
		$1100 k/he	$250 k/he	$500 k/he
Coke ovens (no dephenolization)				
Low flow	1st	9.5	3.8	1.9
	2nd	360	144	72
	3rd	5×10^3	2×10^3	1×10^3
	Shutdown	4×10^6	2×10^6	9×10^5
Phenol manufacturing				
Low flow	1st	6.0	2.4	1.2
	2nd	560	220	110
	Shutdown	2×10^5	1×10^5	5×10^4
Phenolic resins				
Low flow	1st	5.8	2.3	1.2
	2nd	570	230	110
	Shutdown	5×10^4	2×10^4	1×10^4

[a] k/he = 1000 per health effect

three slopes for cost-effectiveness: 1, 0.5, and 0.1, with a value of life of $500,000 per health effect. In only two cases, both on low-flow sites, is the range of slope unity: first level control for phenol manufacturing and for phenolic resins. The rest are above unity slope, and, taking into account indirect investment expenses (slope -0.5) and industry shutdown (slope $= -0.1$), would not be considered, This approach provides, as it did for cadmium, some perspective on cancer control. However, one would not apply a control assuming a value of life less than $500,000 per health effect. If one attempts to use mid-range health estimates, the process breaks down. It is extremely sensitive to the uncertainty in health impact.

II. DESIGN APPROACHES

A design approach, as previously discussed, is one which takes into account the type of control technology applied. We consider two types, BPT and BAT, which set different conditions for that technology.

A. Best Practical Control Technology

None of the mid-range estimates for cost-effectiveness shown in Table 60 are reasonable given a BPT approach. For the worst case, there are four industries within $1 million per health effect; coke ovens with no dephenolization, phenol manufacturing, petrochemical, and phenolic resins. There are five, including oil refineries, within $10 million per health effect. These values are only for the first level of control at low-flow sites. Thus, control of all plants on an average-flow basis is not cost-effective. Only first-level control on low-flow sites for the industries listed attains reasonable cost-effectiveness (worst case) for the BPT approach.

B. Best Available Control Technology

The range of cost-effectiveness values for BAT is 5.6×10^7 per health effect to 1.4×10^{16} per health effect (worst case). For low-flow sites, the range is still from 5.6×10^7 per health effect to 1.4×10^{13} per health effect. Thus, none of the controls are cost-effective under any conditions.

Chapter 34

SUMMARY — PHENOL CANCER RISK

The public . . . demands certainties; it must be told definitely
and a bit raucously that this is true and that is false.
But there are no certainties.

Henry Louis Mencken
Prejudices, First Series (1919), Ch. 3

It is humbling to realize that one is not, nor can ever be, *absolutely certain* of the risk of a substance. However, we may say that in light of our current information and state of knowledge, phenol does not appear to pose a great risk as a water pollutant. Without measurable noncarcinogenic chronic effects and with a minimal cancer effect based upon scant data, phenol is a lesser problem than cadmium. Even when we include chlorinated phenols in drinking water and assume a 100 times higher potency and 100% conversion, cost-effectiveness for potential health effects reduction is excessive.

Many of the results obtained for phenol parallel those for cadmium: zero risk has no meaning under any scenario; as low as can be controlled, BPT, and BAT design approaches are all unreasonable. Because the detection limit is so low, the as low as can be measured approach only has meaning if we use a health effect threshold, rather than a detection limit. The health effect threshold yields cost-effectiveness figures that are a bit more reasonable, although still above $1 million per health effect. Nevertheless, use of this threshold could serve as a technique for setting *de minimus* levels.

As low as background, potentially useful for cadmium as a tool for discriminating between average- and low-flow cases, does not work for phenol where background levels are high in both average- and low-flow cases.

Risk comparisons for phenol provide the same conclusions as the as low as can be controlled approach. This is because one needs maximum control at the second level to make risk comparable to other societal risks. A set value for dollars to be spent may be practical for the larger phenol-emitting industries with economies of scale even though the cost-effectiveness is still quite high. This is a situation-specific approach.

Employing marginal cost of risk reduction as a criterion helps establish the need for standards for certain phenol-emitting industries, just as it did for cadmium. Tax incentives do not appear effective for phenol. Risk/cost/benefit balancing may provide perspective for a few industries, but considers neither indirect cost nor effects of industry shutdown.

Part X

OVERALL EVALUATION — ALTERNATE APPROACHES TO STANDARD SETTING

The artist may be well advised to keep his work to himself
till it is completed, because no one can readily help him
or advise him with it . . . but the scientist is wiser not
to withhold a single finding or a single
conjecture from publicity.

Johann Wolfgang von Goethe (1749—1832)
Essay on Experimentation

All things considered, this may be the best policy for regulators and regulatees in most situations. Judicious determinations in this regard, however, are essential as it is of no benefit to panic the public or media. In this analysis, we have attempted to make clear the limitations of the study and the possible range of error as well as our assumptions and their bases.

Since we have investigated only two substances, cadmium and phenol, and only water pathways, we have cautiously generalized on the whole array of standard-setting problems. Nevertheless, several conclusions of a general nature seem obvious. These conclusions (which are repeated from the Introduction), an evaluation of the methods, and our overall impression of the value of the investigation appear in this part.

In discussing the utility of various methods of standard setting for the two pollutants, we have not addressed several factors. These include the impact on wildlife other than aquatic organisms and the use of water in agriculture. In the former case, we have assumed instead that levels of cadmium or phenol adequate to protect man also prevent significant population reductions in terrestrial and avian wildlife. Some finfish and, to a greater extent, shellfish bioaccumulate cadmium; an indirect impact might be the curtailment of harvesting of these species. Costs are not related to specific sites and plants, but this could represent a quantifiable cost of continuing to pollute.

Although this relationship between cleanup or prevention of high cadmium water residues and the benefits of seafood production cannot be readily ascertained, it should be weighed for both new and existing plants in setting risk levels for standards.

As noted, cadmium, a heavy metal, can accumulate through agricultural pathways to food. Contaminated water used for irrigation can contribute to the amount reaching man via this pathway. However, we did not investigate for this use of water either alternative water sources or the impact of cadmium absorbed by crops. Finally, we did not explore in depth the indirect ripple effect throughout the economy of increased control costs and industry closure.

Values used are strictly to illustrate the impact of methods. We explicitly avoided value judgments or exact levels.

Table 69 summarizes the effectiveness of the various methods of setting risk levels. In the following chapters, these are examined in detail.

Table 69
SUMMARY OF THE EFFECTIVENESS OF THE METHODS OF SETTING RISK LEVELS FOR THE CASES INVESTIGATED

Method	Chronic risks (cadmium)	Cancer risk (cadmium and phenol)
Zero risk		
Local	Useful	No meaning for existing cases
Global	Useful, but more limited	No meaning even for new cases
As low as can be measured		
Detection limit	Not useful in the cases studied	Odor limit might be effective practical limit for phenol
Health impact	Useful as a pathway analysis limit	Useful as a pathway analysis limit
	Useful for establishing standard	Useful for establishing *de minimus* levels
As low as can be controlled	Often only buys margins of safety	Generally found to be cost-ineffective
As low as background	Useful as a benchmark	Useful as a benchmark
	Can be used for standard setting, but can be cost-ineffective	Can be used for standard setting, but can be cost-ineffective
A set value for risk reduction	Will not work for highly nonlinear or no-effect level type risk	Possibly useful
		Could be based on control performance
		Should also consider residual risk level
Marginal cost of risk reduction	Useful	Useful
	Permits determination of where to use resources most effectively	Permits determination of where to use resources most effectively
	Limitations: as under cancer risk plus it does not resolve margin of safety and type of standard	Limitations: risk and cost uncertainties, need to assign value to health effects, does not consider equity
Tax incentives		
1. Based on cost of control	Feasible in some cases where tax lower than cost of value of technology	Not feasible
	Means of reducing inequity by requiring polluter to pay	Uncertainty in health effect magnified in taxation rate
	Incentive is to require cleanup after pollution	
	Unless real costs or real health impact measurable arbitrary tax rate will result	
2. Based on value assigned a theoretical health effect	Not feasible	Not feasible
		Uncertainty in health effect magnified in taxation rate
Risk cost-benefit balancing	Not feasible	Useful for providing perspective on whether control should be implemented
	Risk estimated large and would dwarf benefits when value of a life factored	Not feasible
	Balance dependent upon uncertainty in health impact estimates	For setting levels of risk for a standard
Design approaches	Application on a situational level may be effective	In general, the generic approach is counter-productive
1. BAT	Cost-ineffective	Cost-ineffective

Table 69 (continued)
SUMMARY OF THE EFFECTIVENESS OF THE METHODS OF SETTING RISK LEVELS FOR THE CASES INVESTIGATED

Method	Chronic risks (cadmium)	Cancer risk (cadmium and phenol)
2. BPT	Cost-effective First level of control only (worst case situation)	Cost-ineffective under any scenario of projected health effect
3. Other		Criteria for situational application should be expanded to include all those used in setting standards
Comparisons	Useful as benchmarks	Useful as benchmarks Useful when compared with similar cancer risks
Arbitrary risk numbers	Not a useful concept when thresholds exist	Could be useful in establishing a *de minimus* level, but not an ''acceptable'' level
A set value for dollars to be spent	Useful if there is a source resource limitation or value limit to primary products, at least, as an economic checkpoint	Useful if there is a source resource limitation or value limit to primary products, at least, as an economic checkpoint

Chapter 35

OVERALL EVALUATION — RISK AVERSION METHODS

Men will find that they can prepare with mutual aid
far more easily what they need, and avoid far more
easily the perils which beset them on all
sides, by united forces.

Benedict Spinoza
Ethics (1677), pt. III, 29, proposition 35

Risk from certain types of pollution may be effectively combated on the societal level. To do so, society must first determine whether and to what extent the presence of an unwanted substance should be reduced by considering the degree to which this risk should be avoided.

This chapter provides an overview of the risk aversion methods as applied to the risks we have been considering. These are summarized in Table 70, a subset of Table 69.

It is obvious from this analysis that the assessment of chronic and cancer risks are best served by entirely different approaches. For cadmium, the impact of high residues of this heavy metal is measurable and a no observable effect level is apparent. If one assumes there is no cancer threshold, very low exposure levels should be considered. In our analysis, the two substances studied are not potent carcinogens. Perhaps all that can be said is that they may be weakly carcinogenic, possibly synergistic or co-carcinogenic, or not harmful at all in this sense. Certainly, relative potency levels can be ascertained. Phenol is less potent than cadmium, assuming both are carcinogenic in laboratory animals at very high doses. This relative potency, along with degree of exposure, provides a means of focusing resources and regulatory attention.

I. ZERO RISK CRITERION — LOCAL AND GLOBAL

When dealing with substances which have a no observable effect condition (NOEL) or are highly nonlinear over the dose-effect range, such as chronic cadmium risks, a zero risk criterion can be quite meaningful. As long as the effect threshold is high enough that available controls result in exposures below the threshold, one can achieve zero risk. This does not mean that attaining zero risk is useful in every situation. For example, under conditions for control of captive electroplating plants of low-flow streams, the exercise of any level of control implies shutdown. Conversely, the number of effects without controls is very high due to nonlinearity of the exposure-effect relationship.

Two other aspects are immediately apparent. First, when exercising control strategies, the initial level of control that brings exposure below the NOEL is the only one which actually reduces risk. All succeeding levels provide increasing margins of safety but at a price. It is evident that selecting a margin of safety beforehand or "across-the-board" tends to mask this cost. We do not argue against the concept of margins of safety. However, it is evident that the cost of different degrees of margin of safety can be calculated and that decisions on the level to be implemented should be made on a visible, rational basis. Our analysis of margins of safety is not geared to the most sensitive members of the population nor does it consider the nature and degree of sensitivities. We did not investigate possible short-term excursions of standards. Margins of safety are, of course, throught to provide some degree of protection. If they can be ascertained, one can determine even the cost-effectiveness of various degrees of safety margin.

We conclude that, before selecting a margin of safety, its cost and effectiveness should be determined, not the other way around. Again, the results are situation-oriented; and, while margins of safety are important, one should know the price paid to achieve them.

Table 70
SUMMARY OF THE EFFECTIVENESS OF RISK AVERSION METHODS OF SETTING RISK LEVELS FOR THE CASES INVESTIGATED

Method	Chronic risks (cadmium)	Cancer risk (cadmium and phenol)
Zero risk		
Local	Useful	No meaning for existing cases
Global	Useful, but more limited	No meaning even for new cases
As low as can be measured		
Detection limit	Not useful in the cases studied	Odor limit might be effective practical limit for phenol
	Useful as a pathway analysis limit	Useful as a pathway analysis limit
Health impact	Useful for establishing standard	Useful for establishing *de minimus* levels
As low as can be controlled	Often only buys margins of safety	Generally found to be cost-ineffective
As low as background	Useful as a benchmark	Useful as a benchmark
	Can be used for standard setting, but can be cost-ineffective	Can be used for standard setting, but can be cost-ineffective

The second aspect is related to the global zero risk situation. One cannot establish a zero risk level or a margin of safety to a substance with a nonlinear dose-effect relationship without considering total exposure from all sources. This becomes evident when investigating cadmium pollution where the major human exposures are via food and smoking. In a sense, it is folly to set a zero risk standard in one media if the level in another violates it. Moreover, it may be more cost-effective to reduce risk in another environmental medium than the one in question. This calls for an across-media approach. The Environmental Protection Agency was originally set up in 1970 to use this approach. However, because the EPA legal authorities and organizational structures have all been media-oriented, such across-media attention has been sparse. Nevertheless, it should be possible to make this type of analysis, even though standards might have to be established on a media-by-media basis. In fact, EPA has now established an Office of Toxic Integration to synchronize the approaches in various media.

These conclusions do not apply where no effects levels are absent and approximately linear dose-effect relationships exist. Zero risk is a totally meaningless concept in this context. This does not mean, however, that one should not analyze risks to very low levels, but that standards set on the basis of its elimination (zero or very low levels) are meaningless, except in rhetoric.

Moreover, the concept of applying margins of safety to cancer through the use of conservative estimates (in the public health protection sense) is conducive to the same problems as those encountered when setting margins of safety. Conservatisms should be applied in terms of risk differences at the end of the analysis, not when the risk estimate is made. The ranges of uncertainty must be kept open and visible to the end, so that one can see how much is being paid for this conservatism. It is evident from the analyses in Parts VII, VIII, and IX, that maximum estimates provide more cost-effective results for the same situation. Conversely, if the substance is not carcinogenic, the whole process is meaningless.

Certainly a degree of conservatism is desirable, but one should know both exactly what is being purchased and how much it costs. In addition, even for the same substance, the degree should probably not be constant across all situations. It is quite evident that lesser margins of safety are required where exposure pathways are small or negligible. Our argument is not against conservatism, but against blanket application without regard to impact.

The assignment of health impact to a carcinogen via the dose-effect relationship was the most critical decision in the analyses in Chapters 19 and 20. These critical judgments should be made entirely visible and not left to scientists or to regulatory procedure.

II. AS LOW AS CAN BE MEASURED

The as low as can be measured concept is not a reasonable criterion as a detection limit for establishing standards, at least not for the cases evaluated. At best, it provides a lower limit for pathway analysis. Since detection methods are always improving, levels based upon detection capabilities are subject to change.

The precision and accuracy of field measurements may well be limiting situations for implementing and enforcing standards. Moreover, inability to discriminate between discharge background substances in the field may not only make enforcement difficult but may prevent validation of models upon which standards are to be based, e,g., the river flow model in Chapter 11.

In this sense, as low as can be measured is a limiting concept, not useful in making judgments on levels of standards. It is restrictive, not augmenting.

III. AS LOW AS CAN BE MEASURED — HEALTH IMPACT MODELS

This concept is separated from the previous case, since it does not involve direct measurement, but rather the results of applying models. In this concept one does not "count the bodies" by direct observation of health impact through cause and effect relationships. Rather, models are used to estimate the health impact based upon available information, and then the models are used to establish a *de minimus* level low enough to disregard. Such a level is set arbitrarily, but provides a basis for finding a level which is a very small percentage of the *variation* in the least sensitive, critical model parameter.

In this analysis, we selected an arbitrary level of 0.1 health effect in the total U.S. population (whether or not exposed). We could have selected any other level; it is used only for illustration. On this basis, even for our worst case cancer estimates, plants in four (4) cadmium effluent industries and three (3) phenol effluent industries could have been immediately set aside from further consideration if they were not on low-flow sites.

Note that the *de minimus* concept is based upon the health impact of the model. It does not use built-in mechanisms to mask health effects. Such built-in mechanisms, e.g., dose-rate reduction at low doses, as long as they cannot be proven or disproven, obscure the key decision of determining the level of health impact low enough to ignore after calculation. Both potency and exposure potential are used in estimating risks. Thus, the condition is situation, not substance, oriented. Impact is always determined, then disregarded if small. As one acquires information, the status of a situation may easily change.

The *de minimus* concept can eliminate considerable effort, but does not in itself help in setting levels above the threshold. It may, however, be of use when applied in conjunction with other methods. If one could set a level across-the-board (as could a range of levels for different purposes), it could, indeed, conserve resources without contributing to increased risk.

IV. AS LOW AS CAN BE CONTROLLED

As low as can be controlled implies maximum control in all situations. Unless coupled with other methods, it has little meaning. If used with *de minimus* levels, as described above, it can be useful, but in most instances is very cost-ineffective. It may run well into the tens of millions (or greater) dollars per health effect reduced for cancer or it may simply buy increased margins of safety. It limits control to the extent that shutdown is not necessarily implied. It does, however, become a limiting concept in a global situation, where natural or other sources of exposure dominate.

As low as can be controlled is an across-the-board application and differs from some applications of Best Available Technology (BAT). For example, in Section 112 of the Clean Air Act, BAT is only applied if a substance is shown to be a hazardous air pollutant as defined in the Clean Air Act.

In the cases studied, the as low as can be controlled concept provides some upper limit considerations, but where risk existed, it was very cost-ineffective with or without *de minimus* limits.

V. AS LOW AS BACKGROUND

As low as background can be defined in three ways:

1.   A level which cannot be discriminated from background
2.   Some measurable fraction of the natural variation in background
3.   A level equal to and in addition to background

In the first definition, the concept reduces to as low as can be measured. Using the second, a useful criterion might be developed on occasion. This second approach is based upon equity considerations since everybody does not receive the same background exposure and a small portion of the uncontrolled inequity might well be redistributed. Depending upon the level chosen for selecting percentage of variation, e.g., on standard deviation or a fraction of one, this may have utility in establishing a standard.

The third definition assumes that not only is the background level safe, because it cannot be controlled, but that twice that amount is also safe. This is at best arbitrary and is only meaningful if a linear dose-effect model is postulated or if the presumed safe amount is below a no-effect level.

Based upon ambient water levels for noncarcinogenic chronic cadmium risks, we set levels which, although arbitrary, are useful as benchmarks. We similarly found levels for cancer risk from the same industries. In these situations, lifetime risk was low, but calculable (4×10^{-6} lifetime — worst estimate, 4×10^{-8} — mid-range estimate, zero).

Chapter 36

OVERALL EVALUATION — ACCEPTABLE RISK METHODS

Much Madness is divinest Sense —
To a discerning Eye —
Much Sense — the starkest Madness —
'Tis the Majority
In this, as All, prevail —
Assent — and you are sane —
Demur — you're straightway dangerous —
And handled with a Chain

Emily Dickinson
No. 435 (c. 1862)

Since the greatest uncertainties in analysis were from health impact and associated risk estimates, the use of acceptable risk methods depends upon choice and range of risk estimates. This is a key decision; if the range is too wide, one may get inconclusive results; if too narrow, incorrect values. If conservative bias is already included in estimates, this may result in entirely slanted and obscured conclusions. In this latter case, wide uncertainty is thus coupled with subjective value judgments of acceptability. Although one cannot get precise answers under these conditions, useful information may still be derived. Table 71 summarizes our evaluation.

I. COMPARABLE RISKS

The first step for a risk comparison is to determine the risks against which the effluent risk is to be compared. A particular risk at a given level could be compared on two aspects simultaneously: (1) the risk level and, (2) the range of uncertainty in risk estimates. A particular reference risk might be selected because of similarity to cadmium and phenol in terms of consequences and the uncertainty range in the estimate for the reference. For example, if the reference risk estimate ranges from zero to a worst case level similar to that of cadmium or phenol, it would then at least have the same range of uncertainty.

If such reference risks are accepted elsewhere in society, they provide benchmarks for similar levels of acceptance in our standard setting. Choice of acceptable references, then, is the key judgment. They should be used primarily as benchmarks, since in some instances a risk might be easily eliminated while another risk of the same magnitude may be very difficult to control. Therefore, benchmarks are not firm rules but guides providing us an understandable perspective. Such perspective is not provided by numerical values, and may be extremely important in presenting risk estimates.

A selection of a set of references is a judgment on the underlying arbitrary risk numbers that they represent. Therefore, most of this discussion is also applicable to arbitrary risk numbers.

II. ARBITRARY RISK NUMBERS

Although one may reference arbitrary risk numbers to benchmarks, to gain perspective, raw risk numbers may also be set. There are two decisions or judgments involved. The first judgment is what level, e.g., 10^{-5}, 10^{-6}, 10^{-7}, of health effects per year or per lifetime should be used as a reference. Selection can be difficult. The second judgment is which risk estimate to use for comparison with the reference. As shown in Table 57, our mid-range estimate

Table 71
SUMMARY OF THE EFFECTIVENESS OF ACCEPTABLE RISK METHODS OF SETTING RISK LEVELS FOR THE CASES INVESTIGATED

Method	Chronic risks (cadmium)	Cancer risk (cancer and phenol)
Comparisons	Useful as benchmarks	Useful as benchmarks; useful when compared with similar cancer risks
Arbitrary risk numbers	Not a useful concept when thresholds exist	Could be useful in establishing a *de minimus* level, but not an "acceptable level"
A set value for dollars to be spent	Useful if there is a source re-source limitation or value limit to primary products, at least, as an economic checkpoint	Useful if there is a source re-source limitation or value limit to primary products, at least, as an economic checkpoint
A set value for risk reduction	Will not work for highly nonlin-ear or no effect level type risk	Possibly useful. Could be based on control performance. Should also consider residual risk level

for cadmium-caused cancer requires no controls at 10^{-6} lifetime risk but the worst case estimate requires control at all low-flow sites at this level. Table 66 showed similar results for phenol.

The use of benchmarks, as discussed, allows one to pick references with somewhat similar uncertainty ranges to reduce the impact of risk estimation uncertainty. Use of any single number fails in this respect, since it would not take into account the confidence in estimates, the ranges of uncertainty, the bias in estimating the risks, etc. Nevertheless, such risk numbers might be useful as limit values for *de minimus* conditions and for *action levels** at high risk levels. If a risk estimate is orders of magnitude above or below the number, decisions might be made. Usefulness depends entirely on the accuracy of the risk estimate. The lower the number used, the less likely is an accurate estimate.

Since the risk estimate was the most uncertain parameter in this analysis, it was difficult to use arbitrary risk numbers meaningfully, unless the ranges of health impact estimates (the total ranges of uncertainty) were above or below reference levels. Once again, the situation determines the suitability of the method.

III. A SET VALUE FOR DOLLARS TO BE SPENT

We assume that when there is a fixed dollar amount available for control, these dollars are used to cost-effectively reduce risk, via marginal cost-effectiveness, up to the amount available. This approach is better if coupled with the value of technology as we did in evaluating approaches to standards for each source of risk. You will recall that we used levels of 1 or 10% of the value of technology. In most respects, this approach is primarily economic in that it relates the amount to be spent for risk reduction directly to productivity and cost consequences to the public. It is not applicable to industries with small plants and high control costs as in the electroplating industry.

If one assumes funds will be spent most cost-effectively (via marginal cost of risk-reduction), this approach limits spending and ties it indirectly to benefits. The method requires an economic value judgment, but only involves risk in terms of the risk reduction afforded. It becomes a useful rule of thumb, once we determine a risk level, to see the ratio of control cost to industry value. Too high or too low a proportion would raise questions.

---

*   The level, generally above a margin of safety, where regulatory action will be legally defensible.

IV. A SET VALUE FOR RISK REDUCTION

This approach is based upon risk reduced, not residual levels of risk as in previous methods. Since the risk level is different in each case, we might use a percentage risk reduction. Note that this does not work for a highly nonlinear risk or risk where there is a no observable effect level type risk. It may, however, make sense with cancer risk, and could be expressed either in terms of percentage of health effects reduced or the reduction (decontamination) factor of controls that must be applied. In fact, judgment could be based upon the performance of controls and related to a choice among BPT, BAT, etc. In general, it makes no sense to address the risk reduced without referring to the residual risk level.

Chapter 37

OVERALL EVALUATION — ECONOMIC AND DESIGN APPROACHES

*Equity is a roguish thing. For Law we have a **measure**, know
what to trust to; Equity is according to the conscience of
him that is Chancellor, and as that is larger or
narrower, so is Equity. 'Tis all one as if they
should make the standard for the measure we call a
'foot' a Chancellor's foot; what an uncertain
measure would this be! One Chancellor has a
long foot, another a short foot, a third
an indifferent foot. 'Tis the same thing in
the Chancellor's conscience.*

John Selden
Table Talk (1689) Equity

Equitable administration of government is still beset with problems. This is as true with regard to costs as it is to risks. This chapter provides an overall evaluation of economic and design approaches. Table 72, another subset of Table 69, summarizes these methods of setting risk levels.

When considering the type of standard which can be set, those responsible for implementation and enforcement usually focus on those which can be easily applied on the same basis to all parties and for which noncompliance is easily measured. Conversely, regulatory decision makers, especially in the climate of today, are looking for effective ways of regulating at low cost, both for themselves (administratively) and for the regulated industries.

I. ECONOMIC APPROACHES

Standards set across-the-board, independent of the local situation, often result from the first approach mentioned above. A concentration or amount standard, set at the discharge point independent of the local dilution capability, is an example. It is extremely cost-ineffective, as exemplified in the cases investigated here, where problems are on low-flow river sites and regulation at other sites is not effective. The entire industry would bear the cost equally in this situation since all must apply controls upon pipe discharge.

Regulations which take into account local conditions are more cost-effective, but require only certain industries to bear expenses. Standards based upon downstream concentrations are an example. The model used may be applied to all sites equally, but only those on low-flow rivers must install controls. In addition, some plants, because of the effect of location on human exposure have higher control costs than competitors. These sites can be avoided when planning a new installation, but those already there would pay a penalty.

These existing low-flow industries may, on the other hand, be subsidized if water quality or drinking water standards are adopted. This would shift the control burden from polluter to water user and is, of course, inequitable.

Some have suggested various taxation schemes to redress inequity. These are somewhat different from either "the polluter pays" concept or the use of taxes to motivate installation of efficient controls. For example, under one of these schemes, controls for cadmium and phenol could be installed only at low-flow sites, and expenditures subsidized by an equitable tax on all producers based upon production. In this case, the control costs at a few sites is borne equally by all parts of the industry. Unfortunately, this approach has not been used to date so we have no actual examples.

Table 72
SUMMARY OF THE EFFECTIVENESS OF ECONOMIC AND DESIGN METHODS OF SETTING RISK LEVELS FOR THE CASES INVESTIGATED

Method	Chronic risk (cadmium)	Cancer risk (cadmium and phenol)
Marginal cost of risk reduction	Useful. Permits determination of where to use resources most effectively. *Limitations:* as under ''Cancer risk'' plus it does not resolve margin of safety and type of standard	Useful. Permits determination of where to use resources most effectively. *Limitations:* risk and cost uncertainties need to assign value to health effects, does not consider equity
Tax incentives		
1. Based on cost of control	Feasible in some cases where tax lower than cost of value of technology. Means of reducing inequity by requiring polluter to pay. Incentive is to require clean up after pollution. Unless real costs or real health impact measurable arbitrary tax rate will result	Not feasible. Uncertainty in health effect magnified in taxation rate
2. Based on value assigned a theoretical health effect	Not feasible	Not feasible. Uncertainty in health effect magnified in taxation rate
Risk cost-benefit balancing	Not feasible. Risk estimated large and would dwarf benefits when value of a life factored. Balance dependent upon uncertainty in health impact estimates	Useful for providing perspective on whether control should be implemented. Not feasible for setting levels of risk for a standard
Design approaches	Application on a situational level may be effective	In general, the generic approach is counter-productive
1. BAT	Cost-ineffective	Cost-ineffective
2. BPT	Cost-effective. First level of control only (worst case situation)	Cost-ineffective under any scenario of projected health effect
3. Other		Criteria for situational application should be expanded to include all those used in setting standards

Another alternative, where taxation can be employed, is to use either water quality or drinking water standards, charging the cost of control to pollution industries through specific effluent tax rates. This does affect many different pollutants from multiple sources; however, identifying the source and allocating responsibility is difficult. In spite of this, it may be well worth further exploration. Other taxation approaches are primarily motivational rather than equity preserving.

A. Marginal Cost of Risk Reduction

With certain limitations, marginal cost of risk reduction is a very useful concept. These limitations are primarily due to the multiplicative relationship in the uncertainties in both risk estimates and control costs, and the need (explicit or implied) to put some value on avoiding an adverse health effect.

A major advantage is that resources may be used effectively either for all risks of a similar nature or for control in specific situations. Determining when to stop spending or, alternatively, where to set a limit on dollars per health effect avoided is strictly a judgment.

No regulator wishes to publicly state an explicit value for health effect reduction. However, values spanning $50,000 to $1 million per health effect avoided have been considered, including an EPA Radiation Program "rule of thumb" of $100,000 to $500,000 per health effect. The limits of these ranges provide a first-cut discrimination function. Below the lower limit, implement controls; above, do not; mid-range, bring to bear the total range of approaches to setting levels. Levels may be so far outside the mid-range values that uncertainty in risks has really very little impact.

For chronic noncarcinogenic risks, all first-level controls for mining and smelting and for phosphate fertilizer industries cost less than $1200 per health effect avoided in even the most optimistic health risk case. Those for electroplating shops are below $10,000 per health effect for low-flow sites and go up to $97,000 per health effect for control at all sites (the need for the latter is very questionable). Even ambient and drinking water controls are very cost-effective.

For cadmium cancer impact (shown in Table 50), only the first level of control for mining and smelting on low-flow sites (worst case estimate) results in costs below 10^6 per health effect. All other cases (except for phosphate fertilizer plants) under similar conditions are orders of magnitude higher.

The same is true for control aimed at phenol. Only four industries (Table 60) located on low-flow sites have first-level control cost-effectiveness below 10^6 per health effect (as a worst estimate). None are below $500,000 per health effect. In the cases here, the marginal cost-effectivness was nearly always outside the range of the discrimination function. Those above the range essentially did not warrant control. Those below required a consideration of control; but since they all involved chronic risks, questions of margin of safety and what kind of standards should be set is still open.

Given these limitations, finding marginal cost-effectiveness of risk reduction is still useful. If it is not used to set the level of control, it can at least provide a means of discriminating among extremes. It can only be used to set an acceptable level of risk when one predetermines a specified value for health effects avoided. Even then, one must evaluate the range of risk uncertainties in the final analysis.

This approach does not in itself address equity questions such as who pays, who is at risk, and who is causing the discharge. Even if the polluter pays for the controls to reduce risk, there is still some inequity since residual risks are borne by the general population. This problem of equity is complex, but underlies all of the economic approaches and must be addressed when selecting the type of standard.

B. Tax Incentives

We examined two tax incentive approaches; one for industry to motivate installation of controls and the other for water users who would pay according to the theoretical value of a health effect. For chronic noncarcinogenic cadmium risks, tax incentives based upon the first approach were feasible for effluent and drinking water control costs in some industries (Table 49). Feasibility is defined here as whether the tax would be lower than the value of technology — not really a very good measure. However, since any cost above that of the value of technology would cause industry to shut down, using the value of a health effect prevented was ineffective even when we placed low values on a life.

Alternate water supplies are obtained only at costs which represent substantial percentages of the value of the technology. These costs amount to 23% for drinking water standards and 14% for water quality standards (alternate water use) for the captive electroplating industry. Since this industry would fail if effluent controls were implemented, this is a possible approach. The purpose is not to motivate the industry to install controls, but to allow them to pollute if they choose and then assess them the expense of cleanup.

For potential cancer effects from either cadmium or phenol, incentives through taxation would not motivate installation of controls. As shown in Tables 58 and 67, in only one case, low-flow mining and smelting for cadmium, is the tax rate above or even near the break-even level. Above this point, there is an incentive to apply controls; below it, to pay and pollute. Tax incentives based upon the value of health impact generally are not feasible. Moreover, they are, at least in the cancer cases, completely dominated by the uncertainty in health effect impact. Each order of magnitude in health impact uncertainty reflects directly on the level of taxation by a factor of 10.

We conclude that tax incentives do not motivate pollution reduction in most cases examined. However, it may reduce inequity by eliminating payment for cleanup by water users or at least provide some semblance of equity for making polluter pay for controls. In the latter situation, with neither real costs nor real impact measurable, tax rates would be arbitrary in virtually every instance.

C. Risk/Cost/Benefit Balancing

Risk/cost/benefit balancing does not work where risks are very large, as for chronic cadmium effects. Here, *any* value of a life so inflates the health impact as to totally dominate the benefit side (Figure 28). Of course, this balance depends upon the degree of uncertainty in health impact estimates.

For cancer, risk/cost/benefit balancing provides perspective which helps decide whether to implement controls. However, it may be too insensitive to permit setting a level for a standard. The decision to allow for indirect costs and shutdown values is implied, but how can these be measured? We have used only arbitrary levels for illustrative purposes. Each order of magnitude of uncertainty in health impact changes the slope an order of magnitude. We used only worst case estimates in Tables 59 and 68. When attempting a mid-range estimate, the process fell apart.

II. DESIGN APPROACHES

Across-the-board design approaches, such as BPT or BAT, do not work where dilution reduces ambient levels. However, they can be meaningful under certain conditions, as on low-flow sites in the cadmium mining and smelting and phosphate fertilizer industries, based upon chronic risk. For worst case health estimates, these same industries on low-flow sites become cost-effective for the first level of control only (BPT). BAT is not cost-effective for cancer risk reduction in any instance and only buys an increased margin of safety for chronic risks.

The across-the-board application of BPT results in extremely cost-ineffective controls in the majority of situations, for both chronic or cancer risks. We conclude that generic application of this approach is counter-productive. If true for BPT, it is even more so for BAT.

On the other hand, one can effectively apply controls on a situational basis. Thus, the definition of BPT or BAT depends upon the situation. We have thus far used several general criteria, e.g., first-level control (BPT), maximum control without causing shutdown (BAT); but specific cases might consider any of the following criteria:

- Cost
- Cost as percentage of the value of technology
- Health effects reduced
- Health effects remaining
- Cost-effectiveness of health effect reduction
- Increased margins of safety for maximum dose-effect relationships

In other words, one returns in a circular fashion to the same types of criteria used for establishing standards. Across-the-board application does not make sense and situational application is undefinable without considering the specific case.

Chapter 38

CONCLUSIONS

The country needs and, unless I mistake its temper,
the country demands bold, persistent experimentation.
It is common sense to take a method and try it.
If it fails, admit it frankly and try another.
But above all, try something.

Franklin Delano Roosevelt
Address, Oglethrope University
(May 22, 1932)

We believe the current manner of setting levels of risk in standards by dictating beforehand a standard method to be followed produces less than adequate results. Another approach is needed.

The method for setting levels of risk in standards must fit the nature of the problem. The magnitude and precision of health and environmental impacts, the performance and costs of controls systems, the pathways and extent of exposure, and the direct and indirect value to both society and industry of the technologies causing discharge are all factors in the problem. Because measurement of these factors is often highly nonlinear, if the ranges of uncertainty are of large magnitude, the measurement problems, not the factors themselves, may be the critical parameter. Given this, it may be impossible to predict which methods are useful prior to making at least a partial analysis.

Unless like problems are grouped, each standard setting analysis is situational; methods that work under one set of conditions may be inappropriate to others. *This means that there is no single universally applicable method or sets of methods for setting levels of risk in standards.* In fact, applying a single concept such as Best Available Technology (BAT) before analysis may very well be counter-productive.

This does not mean there are no effective methods for setting standards and establishing acceptable risk levels; only that the situation determines the method, not the reverse. This appears in direct conflict with the general practice of Congress which explicitly states the method to be used without considering the individual situation. The breakdown of the Delaney amendment in the saccharin case, the difficulty in applying BPT and BAT in the Clean Water Act, the Supreme Court reversal of OSHA in the Benzene Case, and the meaning of "unreasonable risk" in TSCA are examples of Congressional attempts at defining methods with subsequent collapse of these systems. From our conclusions, it seems appropriate that Congress provide only the intent of regulation, leaving methodology to the executive agencies. agencies.

For their part, regulatory agencies would, of course, prefer a single method relieving them of all judgments. This is just not possible. Analyses are required for each situation and methods must be chosen to fit the case. When they are made, judgments should be documented and made visible. They should be credible and defensible, never made by procedural dogma. The absence of any pat methodology for resolving a situation may be unsettling to some, but there are several ways of ameliorating the situation.

The first way is procedural. It provides the steps in the analysis, the data to be obtained, the criteria for selecting particular methods, and the manner in which the final judgments should be developed and documented. It also requires a separate analysis from the beginning for each case.

A second way is to prepare an expanded version of this type of study early in the assessment process. Enough information would be developed to identify the factors in a given situation that make certain methods useful and others not. That is, a history of what works in a given situation would guide the choice of method *prior to* the actual analysis. Although it requires considerable development, and may not work under certain extreme conditions, use of this approach ensures that the situation and methods are matched before an in-depth analysis is made. Thus, a preliminary analysis for a particular method would conserve resources by only requiring data necessary to its particular requirements.

A third method somewhat beyond the scope of this study is to provide a menu of the different approaches showing the range of factors and conditions for which a particular method is useful. Thus, when an analysis is undertaken and the factors and conditions for the particular situation are identified, one or more particular methods for setting risk levels is directly selected from the menu by pre-existing criteria. This removes any arbitrariness from the selection of an appropriate method. Of course, many studies such as we have conducted here would be required to develop such a menu.

Note that these comments are directed only to analyses and risk assessments made for the purpose of setting levels of risk in standards. They may or may not apply to other kinds of analyses and risk assessments. Choosing among new technologies, screening for health impact of new substances, and conducting environmental impact analyses are examples of situations where these comments may be inappropriate.

There are times when analytical methodology will not resolve the problem. Where there are major inequities as to sharing the burdens and reaping the benefits and where major dislocations in society are required, political aspects become dominant. This is properly so since the arguments relate to which parts of society gain and which lose.

The electroplating industry is a case in point from this analysis. The imposition of controls on sites where there is less dilution would cause plant closure, since the chronic risks are high, the control costs reasonable, but the value of technology is small on a direct basis. Choice of drinking water controls in this case shift the control cost burden away from polluters to users of water and allow plant operations to continue. The latter may be a consideration since closure can cause major disruptions throughout the country. In this situation, the analytical approach can help frame the questions to be addressed politically, but cannot provide answers. Decisions of this type are difficult to make and practicing politicians are often reluctant to make choices as bluntly as the analysis might present them. Those in the political sphere can often sense many shades of gray that a black and white analysis may miss.

It is important to note that our conclusions are drawn from a limited analysis which explored some areas in depth and others, such as impact on a portion of the biota, not at all.

Moreover, we do not attempt to suggest levels for standards, only the ways one may set these levels. Nevertheless, we believe the evidence adequate to conclude the following:

- The situation determines the method or methods used to establish levels for standards. Conversely, there are no methods universally applicable. Moreover, the pursuit and attempted use of methods *believed* universally applicable is wasteful, cost-ineffective, and often counter-productive.
- In the absence of a universal method, analyzing all risks and alternatives provides a perspective allowing us to apply methods as the situation dictates. This analysis and choice of methods, as well as explanation of the choice, can be a visible, traceable process open to all to follow, concurrently providing credibility and flexibility.
- Full-scale analysis may not be necessary in every situation. Enough clues should exist from the preliminary analysis to provide a first-cut evaluation and an identification of methods which will not and which may work. Further in-depth investigation would follow using the method or methods selected. The analysis is made to fit the situation,

not the reverse. Processes can be developed to enable one to carry out these approaches with minimal arbitrariness.

- Conservatism and margins of safety should not be "built in" to the analysis, but left open until cost is determined. Then, one may decide upon implementation in a visible and defensible manner.

- Methods well-grounded in theory often do not work well in practice. Uncertainties in risk/cost/benefit balancing, for example, were so great as to make this approach worthless for the cases considered here.

- It is possible to use *de minimus* levels to establish a risk below which one need not be concerned. There are many approaches to deriving these, but the one selected should be based upon the expected impact. This means that it is based upon risk, not strategies adopted to mask the risk. Implementing *de minimus* levels can conserve administrative costs for both government and industry.

- Uncertainty in health impact dominates all approaches. However, certain methods, such as the marginal cost of risk reduction and arbitrary risk levels, provide excellent guidance as to whether to adopt control over a wide range of health impact uncertainty. Again, success is situational not methodological.

- Analytical approaches cannot resolve value and ethical issues under any conditions. Although one can frame value judgments and problems of inequity by a particular analytical approach, the difficult judgments must still be made on less than objective grounds.

- Research could result in a better understanding of which methods work in what situations and why, as well as in a generic approach to setting standards in a flexible and credible manner, using minimal resources. The objective must be to resolve issues not to continually revisit them.

ADDENDUM

Since the conclusion of our original study, several models similar to those used here, as well as new data bases for stream flow and other inputs, have been developed or are under development.[171]

GLOSSARY

Absolute Risk	A direct estimate of the probability of an event and its consequences and their relationships.
Acceptable Risk	A level of risk for which a gamble is worth taking, or when the risk is imposed the parties affected are not, or are no longer, apprehensive about the risk.
Acceptance Standard	A standard set at a level where socio-economic factors are balanced against risk. It may be based on either performance or design.
Accuracy	The quality of being free from error. The degree of accuracy is a measure of the uncertainty in identifying the true measure of a quantity at the level of precision of the scale used for the quantity.
Achievement Standard	A standard which describes the achievable level of control which will meet protection requirements. It may be based on either performance or design.
Actual Risk	A function of true probability and consequence of an event.
Acute Effects	Generally refers to the toxic effects of a substance which become manifest after only a short period of exposure of a duration measured in minutes, hours, or days.
ADI	Stands for Acceptable Daily Intake. Permissible level of a particular additive in food as determined by the Food and Drug Administration.
Alternative	One member of a set of options associated with a decision, the decision being limited to a choice of one and only one.
Ambient	A term applied to naturally occurring background amounts of a substance in a particular environmental medium. It may also refer to existing amounts in a medium regardless of source.
Ames Test	A standardized screening test using a mutated strain of salmonella bacteria to determine whether introduction of a given substance causes further mutations in the bacteria. Mutagenesis is believed an indication of carcinogenesis.
Annuity Present Worth Factor (PWF)	A derived factor used in evaluating costs via the present worth method. The PWF is defined by the number of uniform payments at fixed intervals for a given discount rate.
A Priori Information	Prior knowledge about the behavior of the system.

Arbitrary Risk Numbers | A chosen specific value of risk below which risks are not controlled across society.

As Low as Background | A risk aversion method in which risk is reduced to or close to the level of that produced by natural background.

As Low as Can be Controlled | A risk aversion method based on the practicalities of actually reducing risk to the lowest level. It considers the contribution of all controllable factors.

As Low as Can be Measured | A risk aversion method based on reducing the source of risk to the lowest measurable level.

Baseline | A known reference used as a guide for further devlopment activities.

Benefit | (a) An axiological concept representing anything received that causes a net improvement to accrue to the recipient. (b) A result of a specific action that constitutes an increase in the production possibilities or welfare level of society.

Benefit-Cost Ratio | The ratio of total social benefit to total social costs related to a specific activity.

Best Available Technology | Demonstrated practice of best industry process for effluent or risk control.

Best Practicable Technology | Average practice of the best industry processes for effluent or risk control.

Bias Error | Error introduced into analysis because of preference or inclination which inhibits impartial judgment and can skew results.

Biological Modeling | Models of the fate and effects of toxic pollutants in biological systems, involving ecological and metabolic systems.

Blind Study | A study in which the subjects and controls have no knowledge of what is being tested and why.

Bubble Concept | A concept for controlling pollution risk in an area by setting an overall regional risk level and allowing individual sources to trade off pollution within that limit.

Carcinogen | A substance possessing the ability to induce cancer in living organisms.

Catastrophes | Major disruptive events which are either natural or man-made and result in multiple fatalities, injuries, and high property damage.

Causative Event	The beginning in time of an activity that results in particular outcomes of the activity.
Chronic Effects	Generally refers to the toxic effects of a substance which become manifest after prolonged or repeated exposures of a duration measured in weeks, months, or years.
Cohort	A defined test population whose lifetime mortality statistics have been determined.
Consequence Value	The importance a risk agent subjectively attaches to the undesirability of a specific risk consequence.
Conservatism	When referring to risk analyses, the tendency to inject a certain bias into an analysis, e.g., a weighting toward protection of human health.
Control	Given a standard of comparisons or means of verification, that which affords a means of directing performance in the direction of the standard of comparison.
Controllability	A performance measure of a system that describes whether the system will perform as it is directed.
Cost	A result of a specific action that constitutes a decrease in the production possibilities or welfare level of society.
Cost-Benefit Analysis	An attempt to delineate and compare in terms of society as a whole the significant effects, both positive and negative, of a specific action. Generally a number of alternative actions are analyzed, resulting in the selection of the alternative that provides either the largest benefit-cost ratio (total benefit/total cost) or one with a positive ratio at least. If an alternative results in a net benefit less than zero or a benefit-cost ratio less than one, it is deemed socially inefficient and is not carried out.
Cost-Effectiveness Analysis	A term less specific than cost-benefit analysis, usually meaning the selection of the lowest cost alternative that achieves a predetermined level of benefits. Alternatively, the analysis and selection of the path that yields the largest social benefit for a predetermined specified level of social costs. Alternatively, the analysis and selection of the alternative closest to a predetermined marginal cost-effectiveness ratio.
Crude Annual Cost Method	A technique for evaluating industry cost in which the capital cost of equipment, apportioned evenly over the lifetime of the facility, is added to the annual operating costs.
Data Base	Available, relevant raw information about the subject of concern.

Decision Making

A dynamic process of interaction, involving information and judgment among participants who determine a particular policy choice. Decision models are either models of the decision-making process itself, or analytical models (e.g., decision trees, decision matrices) used as aids in arriving at the decisions. Decision theories usually are in relation to the process itself.

Degree of Belief

The strength of one's faith that either empirical or subjective conditions may be valid.

Degree of Uncertainty

That proportion of information about a total system that is unknown in relation to the total information about the system.

Delayed Acute Effects

Delayed death or injury as a result of massive exposure to a toxic agent in a specific event or set of events.

De minimus Level

A legally set level of a pollutant below which one need not be concerned.

Design Standard

A standard which specifies the technical design to be used, i.e., a specified control method, e.g., double liming.

Dilution Ratio

The relationship between the volume of water in a stream and the volume of incoming waste. It can affect the ability of the stream to assimilate waste.

Direct Discharger

A facility which discharges or may discharge pollutants directly into waters of the U.S.

Discount Rate

Based on the principle that a dollar received today is valued more than a dollar received in the future, an interest rate that reflects either the social opportunity cost rate or the social time preference rate, permitting the conversion of benefits and costs into present values for analytical purposes.

Discounted Cash Flow

The use of an appropriate interest rate to discount all cash flows to the same point in time.

Dose-Response Curves

Functional relationship between amount of substance and lethality/morbidity.

Dose-Effect Response

Empirical relationship between amount of substance and health impact.

Double Blind Study

A study in which neither the subjects and controls nor the data gatherers know what is being tested or why.

Economies of Scale

Reductions in mininum average costs that come through increases in the size (scale) of plants and equipment.

Effects/Impact Model	A model to determine the degree of risk reduced by a given reduction in pollutant by a standard.
Effluent	Treated or untreated waste material discharged into the environment. Generally refers to water pollution.
Empirical	Originating in or based on observation or experience.
Environmental Fate	The disposition of a substance in various environmental media, air, water, soil, etc.
Environmental Impact	Impact on the biota and abiotic components of the environment.
Environmental Management	A systematic approach to the planning and control of the environment.
Epidemiology	The study of the causes of diseases by identifying personal and environmental characteristics common to those contracting the disease.
Estimation	The assignment of probability measures to a postulated future event.
Equitable	Fair to all concerned without prejudice, favor or rigor entailing undue hardship.
Equitable Risk	Risk taker stands to benefit directly from the event involved.
Evaluation	Comparison of performance of an activity with the objectives of the activity and assignment of a success measure to that performance.
Event	A particular point in time associated with the beginning or completion of an activity, and possibly accompanied by a statement of the benefit or result attained or to be attained because of the completion of an activity.
Expected Risk	Multiplicative function of probability and consequences estimates of a given event.
Expected Value, Use of	Valuation of an uncertain numerical event by weighting all possible events by their probability of occurrence and averaging.
Expert Judgment	Designating the relevance of opinions of persons well informed in an area for estimates (e.g., forecasts of economic activity).
Exposure (to risk)	The condition of being vulnerable to some degree to a particular outcome of an activity, if that outcome occurs.

Exposure Pathways	Means by which risks are transmitted. The route by which a given population is exposed to a toxic substance, i.e., via drinking water, air, dermal contact, etc.
Expressed Preference	Direct questioning of individual's perception.
Extrapolation Among Species	The act of applying a set of data or an individual test result on one species, under certain conditions and subject to particular dose levels of a toxic substance and application method to another population of the same or different species under perhaps different conditions, dose levels, and application method.
Feedback	The return of performance data to a point permitting comparison with objective data, normally for the purpose of improving performance (goal-seeking feedback), but occasionally to modify the objectives (goal-changing feedback).
Gain-Loss Analysis	Direct: comparison of direct gains and losses; primarily economic. Indirect: comparison of indirect societal gains and losses; primarily qualitative.
Gain-Loss Balance	Result of gain-loss analysis.
Groundwater	The supply of fresh water under the surface of the earth that forms a natural reservoir.
Hazard	Danger, peril, threat, which does not necessarily imply potential for occurrence.
Hazardous Substance	A substance whose effect on man or animals is potentially large but undefined since an exposure pathway may or may not exist. It leads only to risk if an exposure pathway exists.
Health Impact	Acute and chronic impact on human life as a result of exposure to an event.
Heuristic	Describing an operational maxim derived from experience and intuition.
Hierarchy	A partially ordered structure of entities in which every entity but one is successor to at least one other entity, and every entity except the basic entities is a predecessor to at least one other entity.
Human Capital Approach	Value of life based on individual's productivity.
Immediate Acute Effects	Immediate death or injury from a specific event.

Immunosuppression	A suppression of the immune response system of living organisms by which it combats invasion of disease organisms and foreign substances.
Implicit Societal Evaluation	A method of assigning a value of a life by estimating expenditures society makes to save a life.
Implied Preference	Consideration of what society has accepted to date and of what current economic conditions allow.
Indirect Discharger	A facility which introduces or may introduce pollutants into a publicly owned treatment works.
Individual Risk Evaluation	The complex process, conscious or unconscious, whereby an individual accepts a given risk.
Inequitable Gambles (Risks)	Exposure of an individual to a hazard for which the individual receives none of the concomitant benefits.
Insurance and Compensation Approach	A method of assigning a value of a life by determining the amount of life insurance purchased and the probability of death caused by a specific condition or activity.
Involuntary Gamble (Risk)	When the risk imposition is: (1) endogenous, and there are no acceptable alternatives or the knowledge of risk was covertly withheld from the risk agent; (2) exogenous, and there are no acceptable alternatives, or the risk is inequitable or both.
LC_{50} (Lethal Concentration 50% Death Rate)	A calculated concentration of a substance in air (or water in the case of aquatic organisms), exposure to which over a specified length of time is expected to cause the death of 50% of an entire defined experimental population.
LD_{50} (Lethal Dose 50% Death Rate)	A calculated dose of a substance which is expected to cause the death of 50% of an entire defined experimental population.
Limits of Knowledge	Binding of mankind to the specific socio-technical state of the age, i.e., errors, conflicting information, limited resources, and measuring devices and inability to control key parameters.
Linear	Straight line. When the statistical relationship between two variables increases on a direct unit for unit basis, this relationship, when plotted on a chart, will form a straight line.
Logistic Curve	An ''S'' shaped curve which, if plotting dose-effect responses, is linear at low doses, of higher degree at higher doses, and finally saturates at very high doses where the effect in question always occurs.

Logistic Model	A model which assumes dose response follows a logistic curve.
Marginal Cost	The increase in total cost associated with a one unit increase in production or in control.
Managerial Judgments	Judgments which interpret and perhaps modify social value judgments so that societal expression may be implemented and enforced.
Margin of Safety	A factor added to a risk level, generally to a NOEL, for purposes of increasing the probability that a standard based on the resultant level will provide increased protection to the general population and individual members from harmful effects of a given substance.
Masked Event	An event which is hidden by the occurrence of spontaneous and competing events and therefore rarely observed.
Measurable	(a) Capable of being sensed, that which is sensed being convertible to an indication; the indication can be logical, axiological, numerical, or probabilistic. If probabilistic, it is empirical and subjective. (b) Comparable to some unit designated as standard.
Measured Risk Level	The historic, measured, or observed risk associated with a given activity.
Measurement Uncertainty	The absence of information about the specific value of a measurable variable.
Methodology	An open system of procedures.
MGD	Millions of gallons per day. MGD is a measurement of water flow.
Mitigation Strategies	Control strategies to reduce risk after exposure or possible exposure.
Model	An abstraction of reality that is always an approximation to reality.
Modeled Estimates	Heuristic study of similar empirical systems which with modification can be used as a model for the system under analysis.
Monitoring	Periodic or continuous sampling to determine the level of pollution and other activity.
Monitoring of Hazards	A recurrent process of observation, recording, and analysis of products, processes, phenomena, or persons for hazardous events or consequences.

Morbidity	In a demographic sense refers to illness and incapacitation from it.
Mortality	Being subject to death or the rate of deaths in a population.
Multi-hit Models	Dose response models which assume more than one exposure to a toxic material is necessary before effects are manifested.
Multiple Correlation	Statistical technique used to determine the existence of relationships among several variables.
Multi-stage Models	Dose response models which assume there are a given number of biological stages through which the ingested material must pass, e.g., metabolism, covalent binding, DNA repair, etc., without being deactivated before manifestation of the effect in question is possible.
Mutagen	A substance possessing the ability to induce heritable mutations in living organisms.
NOEL	Stands for no observable effect level, the amount of a substance which, when administered to laboratory animals, produces no significant changes from an unexposed control group.
Nonpoint Source	A contributing factor to water pollution that cannot be traced to a specific spot; e.g., agricultural fertilizer, runoff, sediment from construction.
NPDES	A National Pollutant Discharge Elimination Systems permit issued under Section 402 of the Clean Water Act.
NSPS	New Source Performance Standards under Section 306 of the Clean Water Act.
Nutritional Mineral	A mineral which is absorbed and utilized in normal metabolism, i.e., calcium and magnesium. Cadmium and some other nonnutrient minerals compete with these minerals for absorption and utilization.
Objective Probability	Assignment of a number between zero and one to an event based on historical trials of occurrence estimation of similar events.
Observable Chronic Effects	Premature death or increased morbidity resulting from long-term exposure to identified substances. Effects may be cumulative and latent.
Oncogenetic	Tumor producing. A distinction between malignant and benign tumors is not attempted.

One-hit Model	A dose response model which assumes response is elicited after a susceptible target has been hit once by a biologically effective unit of dose.
Opinion Survey/Sampling	Any procedure for obtaining by either oral or written interrogation, or by both, the view of any portion of the affected population regarding benefit levels expected, their utility, and/or relative importance. Typically, scientific sampling procedures would be used to maximize (for a given level of effort) the accuracy and precision of the results obtained.
Objective Risk	Estimation of probability and valuation of consequences based on experiments and empirical measurements.
Perceived Risk	Apprehension about the occurrence of an event not necessarily based on knowledge about the objective probability and/or consequence of that event.
Performance Standard	A standard which specifies the performance a system must achieve, i.e., a specified level of risk or a not-to-be-exceeded mean time to failure greater than 100 hr.
Point Source	A stationary location where pollutants are discharged, usually from an industry.
Poisson Distribution	Discontinuous (discrete) distribution with relative frequences at variate values; limiting distribution of a binomial.
Pollution	The presence of matter or energy whose nature, location, or quantity produces undesired environmental effects.
Potency	The quality of being able to cause strong physiological or toxicological effects.
POTW	Publicly Owned Treatment Works.
Precision	The exactness with which a quantity is stated, that is, the number of units into which a measurement scale of that quantity may be meaningfully divided. The number of significant digits is a measure of precision.
Preference	Assignment of rank to items by an agent when the criterion used is utility to the ranking agent.
Present Worth Method (Value)	Time value of future expenses, usually in monetary terms. A technique for evaluating industry cost by adding capital costs to the annuity present worth factor (PWF) and multiplying the total by annual operating costs. It considers the total value of money which can be set aside to later conduct an activity with future costs.

Preventive Strategies	Control strategies to prevent a certain degree of risk to be applied before exposure occurs. They can include standards and other restrictions, modified technology, monitoring, and enforcement.
Primary Treatment	The first stage of waste water treatment; removal of floating debris and solids by screening and sedimentation.
Probability	A numerical property attached to an activity or event whereby the likelihood of its future occurrence is expressed or clarified.
Probability of Occurrence	The probability that a particular event will occur, or will occur in a given interval.
Prospective Study	An epidemiological study in which a healthy group is selected and exposure and effect information gathered and assessed over a period of time.
PSES	Pretreatment Standards for Existing Sources of indirect discharges under Section 307(b) of the Clean Water Act.
PSNS	Pretreatment Standards for New Sources of direct discharges under Section 307(b) and (c) of the Clean Water Act.
Quantification	The assignment of a number to an entity or a method for determining a number to be assigned to an entity.
Range of Values	Evaluation of an uncertain outcome by estimation of maxima and minima for the event.
Regression Analysis	Statistical method developed to investigate the tendency for the expected value of one of two jointly correlated random variables to approach the mean value of its set more closely than the other.
Relative Potency	A measure of the potency of a given substance by comparing that potency with potencies of other chemicals subjected to the same type tests. It can be a means of categorizing by potency both individual chemicals and groups of chemicals.
Relative Risk	An estimate of the relative likelihood of an event in terms of the likelihood of other events of similar magnitude or the comparison of event magnitudes for events with the same likelihood.
Reliability	The probability that the system will perform its required functions under conditions for a specified operating time.
Residual Risk	Risk which remains after an action to reduce the risk has been taken.

Retrospective Geographical Study	An epidemiological study in which past exposure and effects are assessed by geographic region.
Retrospective Study	An epidemiological study of a group with symptoms suspected of being caused by a given toxic agent. Questionnaires, interviews, or records are used to reconstruct the exposure situation.
Revealed Preference	Assumption that society has achieved near optimum balance between risk and benefits associated with any activities and residual risks are acceptable.
Risk	The potential for realization of unwanted, negative consequences of an event. Downside of a ramble.
Risk Acceptance	Willingness of an individual, group, or society to accept a specific level of risk to obtain some gain or benefit.
Risk Agent	See Valuing agent.
Risk Assessment	The total process of quantifying a risk and finding an acceptable level of that risk for an individual, group, or society. It involves both risk determination and risk evaluation.
Risk Averse	Displaying a propensity against taking risks.
Risk Aversive	Acting in a manner to reduce risk.
Risk Aversion	The act of reducing risk.
Risk Balancing	Comparison and equalization of risk.
Risk Brokering	Redistribution of risk without a change in overall risk.
Risk Consequence	The impact to a risk agent of exposure to a risky event.
Risk Conversion Factor	A numerical weight allowing one type of risk to be compared to another type.
Risk Determination	The process of identifying and estimating the magnitude of risk.
Risk Estimation	The process of quantification of the probabilities and consequence values for an identified risk.
Risk Equalization	Spreading of risk on an equitable basis.
Risk Evaluation	The complex process of developing acceptable levels of risk to individuals or society.
Risk Evaluator	A person, group, or institution that seeks to interpret a valuing agent's risk for a particular purpose.

Risk Factors	Conditions which affect the type of consequences, the probability of occurrence and the nature of consequences in respect to risk.
Risk Identification	The observation and recognition of new risk parameters, or new relationships among existing risk parameters, or perception of a change in the magnitude of existing risk parameters.
Risk Management	A systematic assessment/approach to basic organizational type risks.
Risk Preference	Some reference, absolute or relative, against which the acceptability of a similar risk may be measured or related; implies some overall value of risk to society.
Risk Referent	A specific level of risk deemed acceptable by society or a risk evaluator for a specific risk; it is derived from a risk reference.
Risk Reduction	The action of lowering the probability of occurrence and/or the value of a risk consequence, thereby reducing the risk.
Risk Taker	Displaying a willingness to take a chance, whether it be identifiable (personal) or statistical.
Risk Tax	A tax to polluters which increases with the risk involved in producing a product.
Secondary Treatment	Biochemical treatment of wastewater after the primary stage, using bacteria to consume the organic wastes. Use of trickling filters or the activated sludge process removes floating and settleable solids and about 90% of oxygen demanding substances and suspended solids. Disinfection with chlorine is the final stage of secondary treatment.
Screening (hazards)	A process of hazard identification whereby a standardized procedure is applied to classify products, processes, phenomena, or persons with respect to their hazard potential.
Sensitivity Analysis	A method used to examine the operation of a system by measuring the deviation of its nominal behavior due to perturbations in the performance of its components from their nominal values.
Set Value for Dollars to be Spent	A method of establishing an acceptable level of risk by limiting spending for control to a fixed amount.
Set Value for Risk Reduction	A method of establishing an acceptable level of risk by setting values for risk reduction and then comparing what reduction is obtained by imposing each level of control.

Societal Risk Evaluation — The complex process, formal or informal, whereby society accepts risks imposed upon it.

Social Value Judgment — A judgment which balances benefits, costs, and risks while minimizing inequities in benefit-cost balancing.

Stochastic System — A system whose behavior cannot be exactly predicted.

Synergenetic — With identical genotypes, as maternal litter mates in laboratory animals.

Synergism — Production of an effect by two or more agents acting together which is greater in magnitude than the sum of the effects which would be produced individually.

Systemic Effects — Refers to the toxic action of a substance at sites remote from the point of initial contact and requires that absorption and distribution has occurred.

Target of Opportunity — A person exposed to hazardous or toxic substances during the normal course of their occupation or activities. Monitoring can provide dose, exposure, andperhaps effects data.

Technology — The tangible products of the application of scientific knowledge.

Technology Assessment — Systematic determination of the long-term effects of new and existing technologies on humans and the environment.

Technological Risk — Risks created by technology, either new or existing.

Teratogen — A substance capable of causing defects in fetal development.

Tertiary Treatment — Advanced cleaning of waste water that goes beyond the secondary or biological stage. It removes nutrients such as phosphorous and nitrogen and most suspended solids.

Threshold — A discontinuous change of state of a parameter as its measure increases. One condition exists below the discontinuity, and a different one above it.

Toxicity — Inherent ability of a substance to adversely affect living organisms.

Toxic Substance — A substance for which exposure to man or animals results in deleterious effects.

Trans-science — That area which includes political, managerial and social considerations. Judgments in this area are about science but are not science in themselves.

Uncertainty	The absence of information; that which is unknown.
Unobservable Chronic Effects	A contribution to premature death or increased morbidity by the hypothesized synergistic or contributing action of a particular substance.
Unreasonable Risk	A level of risk from a particular cause which an individual or society views as unreasonable.
Valuation	The act of mapping an ordinal scale onto an interval scale (i.e., assigning a numerical measure to each ranked item based on its relative distance from the end points of the interval scale ... assigning an interval scale value to a risk consequence).
Value	A quality quantified on a scale expressing the satisfaction of man's intrinsic wants and desires.
Value of Lives Saved Method	An approach to determine the cost-effectiveness of risk reduction in which a particular value must be placed on a life. There are four methods for arriving at a value: human capital, implicit societal evaluation, insurance and compensation, and the risk approach.
Value of Technology	The value of tangible products resulting from the application of technological knowledge.
Value Judgments	Impact of intrinsic human values on decision making process; either technical, societal, or managerial.
Valuing	The act of assigning a value to a risk consequence.
Valuing Agent	A person or group of persons who evaluates directly the consequence of a risk to which he is subjected. A risk agent.
Voluntary Risk	Imposition of risk involves some motivation for gain; direct relationship between gain and risk exists.
Water Pollution	The addition of enough harmful or objectionable material to damage watr quality.
Water Quality Criteria	The levels of pollutants that affect use of water for drinking, swimming, raising fish, farming, or industrial use.
Water Quality Standard	A management plan that considers (1) what water will be used for, (2) setting levels to protect those uses, (3) implementing and enforcing the water treatment plans, and (4) protecting existing high quality waters.
Worst Case	An event or series of events resulting in the greatest exposure (actual. probable, or possible).

Zero Risk   Attribute of an event with no possible hazard as a result of exposure to or occurrence of event within the limits of knowledge.

REFERENCES

1. **NAS-BEIR,** *The Effects on Populations of Exposures to Low Levels of Ionizing Radiation,* Report of the Advisory Committee on the Biological Effects of Ionizing Radiation, National Academy of Sciences, National Research Council, Washington, D. C., November 1972.
2. Clean Air Act Amendments, 84 Stat., Section 112, 1970, 1676.
3. Staff communication with Environmental Protection Agency, Washington, D.C., June 1980.
4. Federal Water Pollution Control Act, 86 Stat., 1972, 816.
5. Environmental Protection Agency, Office of Research and Development, Research Summary - Industrial Wastewater, EPA-600/8-80-026, Environmental Protection Agency, June, 1980.
6. Clean Water Act, PL 95-217, 1977.
7. **Worsham, J.,** EPA issues steel regulations, *Chicago Tribune,* May 19, 1982.
8. Environmental Protection Agency, Iron and steel manufacturing point source category effluents limitations guidelines, pretreatment standards, and new source performance standards, *Fed. Regist.,* 47(103), May 27, 1982.
9. 15 Changes asked in U.S. water act, *New York Times,* May 28, 1982.
10. **Sugawara, S.,** Court holds EPA to water cleanup program, *The Washington Post,* May 11, 1982.
11. Safe Drinking Water Act, as amended, PL-93-523 Section 1412, 40 FR 11990, 1974.
12. Toxic Substance Control Act, 90 Stat., 1976, 2003.
13. **Rowe, W. D.,** *Anatomy of Risk,* John Wiley & Sons, New York, 1977.
14. *The Washington Post,* October 15, 1979.
15. **Ames, B. N.,** Identifying environmental chemicals causing mutations and cancer, *Science,* 204, 587, 1979.
16. **Cornfield, J.,** Carcinogenic risk assessment, *Science,* November 18, 1977.
17. Assistant Administrator, Environmental Protection Agency Office of Water and Hazardous Substances, Attachment to letter to Chairman, Environmental Protection Agency Environmental Health Advisory Committee, Washington, D. C., April 19, 1977.
18. **Wilson, R.,** Analyzing the daily risks of life, *Technol. Rev.,* 41, February 1979.
19. **Staffa, J. A, and Mehlman, M. A.,** *Innovations in Cancer Risk Assessment (ED01 Study),* Food and Drug Administration National Center for Toxicological Research, Washington, D.C., 1979.
20. **Bunger, B. M., Cook, J. R., and Barrick, M. K.,** *Life Table Methodology for Evaluating Radiation Risk - An Application Based on Occupational Exposures,* Office of Radiation Programs, Environmental Protection Agency, Washington, D.C., 1976.
21. **Vander, A. J.,** *Nutrition, Stress and Toxic Chemicals,* The University of Michigan Press, Ann Arbor, 1971.
22. Food and Drug Administration, Chemical compounds in food producing animals, SOM Docket No. 77 N0026, *Fed. Regist.,* 44(55), 17077, March 20, 1979.
23. **Wilson, E. E.,** The FDA Criteria for Addressing Carcinogens, Testimony submitted to FDA Hearing on Chemical Compounds in Food Producing Animals, SOM Docket No. 77N0026.
24. **Schneiderman,** *Saccharin: Technical Assessment of Risks and Benefits,* Report, National Academy of Sciences, Washington, D.C., 1979.
25. Assessment of Technologies for Determining Cancer Risks from the Environment, Office of Technology Assessment, Congress of the United States, Washington, D.C., June, 1981.
26. **Salsbury, D.,** Comments of the Animal Health Institute on statistical aspects on the SOM document, May 25, 1979.
27. *National Academy of Science - National Research Council and Monograph,* National Academy of Science, Washington, D.C., 1975.
28. **Oser, B. L. and Hall, R. L.,** Criteria employed by the expert panel of FEMA for the gras evaluation of flavouring substances, reprinted from *Food and Drug Toxicol.,* 15(5), 457, 1977.
29. **Saffiotti, U.,** Scientific bases of environmental carcinogenesis and cancer prevention: Developing an interdisciplinary science and facing its ethical implications, *J. Toxicol. Environ. Health,* 2, 1435, 1977.
30. IRLG, Scientific Bases for Identification of Potential Carcinogens and Estimation of Risks, Report by the Work Group on Risk Assessment of the Interagency Regulatory Liaison Group (IRLG) 44 FR 39858, July 6, 1979.
31. **Weinhouse, S.,** Problems in the assessment of human risk of carcinogenesis by chemicals, in *Cold Spring Harbor Conferences on Cell Proliferation,* Vol. IV, *Origins of Human Cancer,* Book C, *Human Risk Assessment,* Hiatt, H. H., Watson, J. D., and Winsten, J. A., Eds., Cold Spring Harbor Laboratory, N.Y., 1977, 1307.
32. **Mack, T. M., Pike, M. C., and Casagrane, J. T.,** Epidemiologic methods for human risk assessment, in *Cold Spring Harbor Conferences on Cell Proliferation,* Vol. IV, *Origins of Human Cancer,* Book C, *Human Risk Assessment,* Hiatt, H. H., Watson, J. D., and Winsten, J. A., Eds., Cold Spring Harbor Laboratory, N.Y., 1977, 1749.

33. **Blot, W. J., Mason, T. J., Hoover, R., and Fravmeni, J. F.,** Cancer by county: etiologic implications, in *Cold Spring Harbor Conferences on Cell Proliferation,* Vol. IV, *Origins of Human Cancer,* Book A, *Incidence of Cancer in Humans,* Hiatt, H. H., Watson, J. D., and Winsten, J. A., Eds., Cold Spring Harbor Laboratory, N.Y., 1977, 21.

34. **Davis, D. L.,** Multiple Risk Assessment as a Preventive Strategy for Public Health, in FDA Symposium on Risk/Benefit Decisions and the Public Health, Staffa, J., Ed., Food and Drug Administration, Washington, D.C., 1979.

35. **Hoel, David G., Gaytor, David, W., Kirchstein, Ruth L., Safiotti, U., and Schneiderman, M. A.,** *Estimations of Risks of Irreversible, Delayed Toxicity,* National Cancer Institute, National Institutes of Health, Bethesda, Md., 1975, 68.

36. **Chand, N. and Hoel, D. G.,** A comparison of models for determining safe levels of environmental agents, in *Reliability and Biometry: Statistical Analysis of Lifelength,* Proschan, F. and Serfling, R. J., Eds., Society for Industrial and Applied Math, Philadelphia, Pa., 1973.

37. **Crump, K. S.,** Fundamental carcinogenic processes and their implications for low dose risk assessment, *Cancer Res.,* 36, 2973, 1976.

38. **Hartley, H. O. and Sielkan, R. J., Jr.,** Estimation of safe dose in carcinogen experiments, *Biometrics,* 33, 1, 1978.

39. **Cornfield, J., Carlberg, F. W., and Van Ryzin, J.,** *Setting of Tolerance on the Basis of Mathematical Treatment of Dose Response Data Extrapolated to Low Doses,* Proc. 1st Int. Toxicological Congr., Plaa, G. L. and Duncan, W. M., Eds., Academic Press, New York, 1978, 143.

40. **Armitage, P. and Doll, R.,** in *Proc. 4th Berkeley Symp. Math. Statist. Prob.,* Vol. 4, University of California Press, Berkeley, 19.

41. **Park, C. N. and Reitz, R. H.,** Mathematical risk estimation procedures for use in establishing virtually safe levels for chemicals, *Health & Environmental Sciences,* Midland, Mich.,

42. **Schroeder, H. A., Nason, A. P., Lipton, I. H., and Belassa, J. J.,** Essential trace metals in man: zinc, relation to environmental cadmium, *J. Chron. Dis.,* 20, 191, 1967.

43. **Gunn, S. A., Gould, T. C., and Anderson, W. A. D.,** Cadmium-induced interstatial cell tumors in rats and mice and their prevention by zinc, *J. Natl. Cancer Inst.,* 31, 745, 1963.

44. **Babich, H. and Stotzky, G.,** Sensitivity of various bacteria, including actinomycetes, and fungi to cadmium and the influence of pH on sensitivity, *Appl. Environ. Microbiol.,* 33, 681, 1977a.

45. *Fed. Regist.,* 44(129), A8664, July 2, 1979.

46. **Slovic, P.,** Psychological determinants of perceived and acceptable risk, *Perceptronics,* 1978.

47. **Otway, H. J.,** Risk assessment and societal choices, *IIASA Research Memorandum,* RM-75-2, 1975.

48. **Lowrance, W. W.,** *Risk Assessment and Catastrophe Management Policies: The Role of Science and Technology,* William Kaufman, Inc., Los Altos, Cal., 1979.

49. **Kates, R. W.,** Risk assessment of environmental hazards, in *Corporate Risk Assessment,* John Wiley & Sons, New York, 1978, 82.

50. **Okrent, David,** Comment on societal risk, *Science,* 208, 372, 1980.

51. Nuclear Regulatory Commission, Safety Goals for Nuclear Power Plants, U.S. NRC NUREG-0880, February 1982, xiii.

52. **Rowe, W. D.,** Risk assessment approaches and methods, in *Society Technology and Risk Assessment,* Conrad, J., Ed., Academic Press, London, 1980, 3.

53. **Linneroth, J.,** *The Evaluation of Life Saving: A Survey,* International Institute for Applied Systems Analysis, Laxenburg, Austria, 1975.

54. American Society for Testing and Materials, Workshop on Regulatory Alternatives and Supplements, Airlie, Virginia, September 16-18, 1982.

55. **Bower, B. T.,** Regional Residuals Environmental Quality Management Modeling, Resources for the Future, Washington, D.C., 1977.

56. **Brannigan, V. M. and Dardis, R.,** Legal and economic criteria for test-based fire risk assessment, in *Fire Risk Assessment,* ASTM Special Technical Publication 762, American Society for Testing and Materials, Philadelphia, Pa.,

57. **Fleischer, M., Sarofim, A. F., Fassett, D. W., Hammond, P., Schacklette, H. T., Nisbet, I. C. T., and Epstein, S.,** Environmental impact of cadmium: a review by the panel on hazardous trace substances, *Environ. Health Perspect.,* 7, 253, 1974.

58. **Baes, C. F., Jr.,** Properties of cadmium, in *Cadmium: The Dissipated Element,* ORNL-NSF-EPO-21, Fulkerson, W., and Goeller, H. E., Eds., Oak Ridge National Laboratory, Oak Ridge, Tenn., 1973, 29.

59. **Friberg, L., Piscator, M., Nordberg, G., and Kjellström, T.,** *Cadmium in the Environment,* CRC Press, Boca Raton, Fl., 1971.

60. **Fulkerson, W. and Goeller, H. E., Eds.,** *Cadmium: The Dissipated Element,* ORNL-NSF-EPO-21, Oak Ridge National Laboratory, Oak Ridge, Tenn., 1973.

61. **Yamagata, N. and Shigematsu, I.,** Cadmium pollution in perspective, *Bull. Inst. Public Health,* 19, 27, 1970.

62. Environmental Protection Agency, STORET computerized data for cadmium in water, Environmental Protection Agency, Washington, D.C.

63. Patterson Associates, Perspectives on Cadmium in the Environment, Report to the National Association of Metal Finishers, Chicago, Ill., July, 1978.

64. **Moll, K., Baum, S., Carpanar, E., Dresch, F., and Wright, R. M.,** *Methods for Determining Acceptable Risks from Cadmium, Asbestos and Other Hazardous Wastes,* Stanford Research Institute, Menlo Park, Calif., 1975.

65. Versar, Inc., Technical and Microeconomic Analysis of Cadmium and its Compounds, EPA, PB-244625, Environmental Protection Agency, Washington, D.C., 1975.

66. **Yost, K. J.,** Some aspects of cadmium flow in the U.S., *Environ. Health Perspect.,* 28, 5, 1979.

67. **Summers, A. O. and Silver, S.,** Microbial transformation of metals, *Ann. Rev. Microbiol.,* 32, 637, 1978.

68. Environmental Protection Agency, Survey of Organic, Metal and Other Inorganic Parameter Concentrations in Selected Region V. Drinking Water Supplies, Water Division and Surveillance and Analyses Division, Chicago, Illinois, April 15, 1975.

69. **Yost, K. J. and Miles, L. J.,** Environmental health assessment for cadmium: a systems approach, *J. Environ. Sci. Health,* A14 4, 285, 311, 1979.

70. **Schroeder, H. A., Nason, A. P., Lipton, I. H., and Belassa, J. J.,** Essential trace metals in man: zinc, relation to environmental cadmium, *J. Chron. Dis.,* 20, 179, 1967.

71. Staff communication, National Cancer Society, New York, 1981.

72. **Krombach, H. and Barthel, J.,** Investigation of a small watercourse accidentally polluted by phenol compounds, *Adv. Water Pollut. Res.,* 1, 191, Discussion, 1, 191, 1962.

73. **Delfino, J. J.,** Contamination of potable ground water supplies in rural areas, in *Drinking Water Quality Enhancement Through Source Protection,* Poiasek, R. P., Ed., Ann Arbor Science Publishers, Michigan, 1977, 275.

74. **Baker, E. L., Landrigan, P. J., Bertozizi, P. E., Field, P. H., Batteyn, B. J., and Skinner, H. G.,** Phenol poisoning due to contaminated drinking water, *Arch. Environ. Health,* 33, 89, 1978.

75. **Rosfjord, R. E., Trattner, R. B., and Cheremisinoff, P. N.,** Phenols: a water pollution control assessment, *Water Sewage Works,* 123, 96, 1976.

76. **Gehm, H., Bergman, J. I., and Beeland, G. V.,** *Handbook of Water Resources and Pollution Control,* Van Nostrand Reinhold, New York, 1976.

77. **Jenkins, C. R.,** A Study of Some Toxic Components in Oil Refinery Effluents, Ph.D. thesis, Oklahoma State University, Stillwater, 1964.

78. National Institute for Occupational Safety and Health, Criteria for a Recommended Standard ... Occupational Exposure to Phenol, U.S. Department of Health, Education, and Welfare, Washington, D.C., No. 76, 196, 1976.

79. **Anon.,** Phenol makers face continuing over capacity, *Chem. Eng. News,* 56, 15, 1978.

80. **Harborne, J. B. and Simmonds, N. W.,** The natural distribution of the phenolic aglycones, in *Biochemistry of Phenolic Compounds,* Harborne, J. B., Ed., Academic Press, New York, 1964, 77.

81. **Lee, G. F. and Jones, R. A.,** Phenols, chlorophenols, and nitrophenols: chemical pollutants of the New York bright, *Priorities for Research,* National Oceanic and Atmospheric Administration, Boulder, Colorado, 71, 191, 1979.

82. **Baird, R. B., Kuo, C. L., Shapiro, J. S., and Yanko, W. A.,** The rate of phenolics in wastewater — determination by direct-injection GLC and Warburg respirometry, *Archiv. Environ. Contam. Toxicol.,* 2, 165, 1974.

83. **Nayar, S. C. and Sylvester, N. D.,** Control of phenol in biological reactors by addition of powered activated carbon, *Water Resour.,* 13, 201, 1979.

84. **Hunter, H. V.,** Origin of organics from artificial contamination, in *Organic Compounds in Aquatic Environments,* Faust, S. J. and Hunter, H. V., Eds., Marcel Dekker, New York, 1971, 51.

85. **Stecher, P. G., Ed.,** *The Merck Index,* Merck & Co., Rahway, N.J., 1969.

86. EPA, Phenol Ambient Water Quality Criteria, Office of Planning and Standards, PB 296 786, Environmental Protection Agency, Washington, D.C., 1979.

87. EPA, Quality Criteria for Water, Environmental Protection Agency, Washington, D.C., 1975.

88. EPA, National Organic Monitoring Survey, Technical Support Division, Office of Water Supply, Environmental Protection Agency, Washington, D.C., 1976.

89. EPA, STORET computerized data for phenol in water, Environmental Protection Agency, Washington, D.C.

90. **Lustre, A. O. and Issenburg, P.,** Phenolic compounds of smoked meat products, *J. Agric. Food Chem.,* 18, 1056, 1970.

91. Bureau of the Census, Standard Metropolitan Statistical Areas, Bureau of the Census, Washington, D.C., 1980.

92. **van der Leeden, F.,** *Water Resources of the World, Selected Statistics,* Water Information Center, Port Washington, New York, 1975.

93. **Visser, S. A., Lamontagno, G., Zoulalian, V., and Tessier, A.,** Bacteria active in the degradation of phenols in polluted waters of the St. Lawrence River, *Arch. Environ. Contamination Toxicol.,* 6, 455, 1977.

94. EPA, National Interim Primary Drinking Water Regulations, 570/9-76-003, Environmental Protection Agency, Washington, D.C., 1975.

95. American Water Resources Association, *Hydrobiology: Bioresources of Shallow Water Environment,* Am. Water Res. Assoc., Proc. Ser., No. 8, National Symposium on Hydrobiology, Miami Beach, Florida, 1980, Weist, W. G., ed., Phillip E. Greeson Publishers, Urbana, Illinois, 1970.

96. Bureau of the Census, *Census of Manufacturers, 1977,* SIC 2435, 2436, 2861, 2865, 2869, 2911, 3312, 3341, 3471, 3479, 3679, Bureau of the Census, Washington, D.C., 1977.

97. Bureau of the Census, Census of Manufacturers, 1972, Bureau of the Census, Washington, D.C., 1972.

98. **Patterson, J. W.,** *Wastewater Treatment Technology,* Ann Arbor Science, Michigan, 1975, chap. 8.

99. **Ruedemann, R. and Deichmann, W. B.,** Blood phenol level after topical application of phenol-containing preparations, *JAMA,* 152, 506, 1953.

100. **Hosler, J., Tschanz, C., Hignite, C. E,, and Azarnoff, D. L.,** Topical application of lindane creams (Quell®) and antipyrine metabolism, *J. Invest. Dermatol.,* 74, 51, 1980.

101. **Hayes, W. J.,** *Toxicology of Pesticides,* Williams & Wilkins, Baltimore, 1975, 521.

102. **Reichenbach-Klinke, H. H.,** Der Phenol Gehalt des Wassers in seiner Auswirkung auf den Fischorganismus, *Arch. Fischwiss,* 16, 1, 1965.

103. Staff communication with Environmental Protection Agency, Washington, D.C.

104. **Fox, M. R. S.,** Nutritional influences on metal toxicity: cadmium as a model toxic element, *Environ. Health Perspect.,* 29, 95, 1979.

105. FDA, Compliance Program Evaluation, FY 75 Total Diet Studies — Adult, U.S. Department of Health, Education, and Welfare, Washington, D.C., 7320.08, 1979a.

106. FDA, Compliance Program Evaluation, FY 75 Total Diet Studies — Infant and Toddler, U.S. Department of Health, Education, and Welfare, Washington, D.C., 7320.33, 1979b.

107. **Johnson, R. D. and Manske, D. D.,** Pesticides and other chemical residues in total diet samples (SI), *Pest. Monit. J.,* 11, 116, 1977.

108. **Johnson, R. D., Manske, D. D., New, D. H., and Podrebarac, D. S.,** Pesticides and other chemical residues in infant and toddler diet samples (I), August 1974 — July 1975, *Pest. Monit. J.,* 13, 87, 1979.

109. **Kostial, K., Kello, D., Jugo, S., Rabar, I., and Maljkovic, T.,** Influence of age on metal metabolism and toxicity, *Environ. Health Perspect.,* 25, 81, 1978.

110. **Ahokas, R. A. and Dilts, P. V.,** Cadmium uptake by the rat embryo as a function of gestational age, *Am. J. Obstet. Gynecol.,* 135, 219, 1980.

111. **Sutou, S., Yamamoto, K., Sendota, H., and Sugiyama, M.,** Toxicity, fertility, teratogenicity, and dominant lethal tests in rats administered cadmium subchronically. II. Fertility, teratogenicity, and dominant lethal tests, *Ecotoxicol. Environ. Safety,* 4, 51, 1980.

112. **Ahokas, R. A., Dilts, P. V., and LaHaye, E. B.,** Cadmium-induced fetal growth retardation: protective effect of excess dietary zinc, *Am. J. Obstet. Gynecol.,* 136, 216, 1980.

113. **Deichmann, W. B. and Witherup, S.,** Phenol Studies. VI. The acute and comparative toxicity of phenol and o-, m-, and p-cresols for experimental animals, *J. Pharmacol.,* 80, 233, 1944.

114. European Inland Fisheries Advisory Commission, Water Quality Criteria for European Freshwater Fish, Report on Mono-hydric Phenols and Inland Fisheries, EIFAC Technical Paper No. 15, Food and Agriculture Organization of the United Nations, Rome, Italy, 1972.

115. **Loomis, T. A.,** *Essentials of Toxicology,* 3rd ed., Lea & Febiger, Philadelphia, 1978.

116. **Casarett, L. J. and Doull, J.,** *Toxicology: The Basic Science of Poisons,* Macmillan, New York, 1975, 149.

117. U.S. Department of Commerce, Toxicology of Metals, Vol. I, NTIS PB-253 991, National Technical Information Service, Washington, D.C., 1979.

118. **Gleason, M. N., Gosselin, R. S., Hodge, H. C., and Smith, R. R.,** *Clinical Toxicology of Commercial Products, Acute Poisoning Home and Farm,* Williams & Wilkins Company, Baltimore, 1969.

119. **Princi, F.,** A study of industrial exposures to cadmium, *J. Ind. Hyg. Toxicol.,* 29, 315, 1947.

120. *Cadmium,* American Industrial Hygiene Association, Hygenic Guide Series, American Industrial Hygiene Association, Akron, Oh., September, 1944.

121. Food and Drug Administration, Food, Drug and Cosmetic Act, 21 U.S.C. Paragraph 30, 1970.

122. **Beliles, R.,** Metals, in *Toxicology: The Basic Science of Poisons,* Casarett, L. J. and Doull, J., Eds., Macmillan, New York, 1975, 354.

123. **Axelsson, B. and Piscator, M.,** Renal damage after prolonged exposure to cadmium poisoned rabbits with special reference to hemolytic anemia, *Arch. Environ. Health,* 12, 374, 1966.

124. **Lewis, G. P., Jusko, W. J., Coughlin, L. L., and Harty, S.,** Cadmium accumulation in man: influence of smoking, occupation, alcoholic habit and disease, *J. Chron. Dis.,* 25, 717, 1972.

125. **Hammer, D. L., Calocci, A. V., Hasselblad, V., Williams, N. E., and Pinkerton, D.,** Cadmium and lead in autopsy tissue, *J. Occ. Med.,* 15(12), December 1973, 956.

126. EPA, Pesticide programs — rebuttal presumption against registration and continued registration of pesticide products containing cadmium, *Fed. Regist.,* 42(206), 56575, 1977.

127. **Suss, R., Kinzel, V., and Scribner, J. D.,** *Cancer Experiments and Concepts,* Springer-Verlag, New York, 1973.

128. **Kjellström, T., Lind, B., Linnman, L., and Elinder, C.-G.,** Variation of cadmium concentration in Swedish wheat and barley, *Archiv. Environ. Health,* 30, 321, 1975.

129. **McKee, J. E. and Wolf, H. W.,** Water Quality Criteria, 2nd ed., The Resources Agency of California, State Water Quality Board, Publication No. 3-A.

130. **Eisler, R.,** Testimony in the Matter of Proposed Toxic Pollutant Effluent Standards for Aldrin-Dieldrin, et al, Federal Water Pollution Control Amendments (307), Docket No. 1., 1974,

131. **Rehwoldt, R. and Karimian-Teherani, D.,** Uptake and effect of cadmium on zebrafish, *Bull. Environ. Contam. Toxicol.,* 15, 1, 1965.

132. **Bertram, P. E. and Hart, B. A.,** Longevity and reproduction of *Daphnia pulex* (de Geer) exposed to cadmium-contaminated food or water, *Environ. Pollut.,* 19, 295, 1979.

133. Environmental Protection Agency, Carcinogen Assessment Group, Problem Oriented Report, Procedure for Calculating Water Quality Criteria, Environmental Protection Agency, Washington, D.C., July 31, 1980.

134. Environmental Protection Agency, Ambient Water Quality Criteria, EPA 440/5-80-021, Environmental Protection Agency, Washington, D.C., 1980.

135. Dow Chemical Company, Toxicological Information on Phenol, Toxicological Research Laboratory, Midland, Michigan, 1975.

136. **Macht, D. I.,** An experimental study of lavage in acute carbolic acid poisoning, *Johns Hopkins Hosp. Bull.,* 26, 98, 1915.

137. National Academy of Sciences, *Drinking Water and Health,* Part II, chap. 6 and chap. 7, A Report of the Safe Drinking Water Committee, Advisory Center on Toxicology Assembly of Life Sciences, and National Research Council, Washington, D.C., 1977.

138. **Deichmann, W. and Oesper, P.,** Ingestion of phenol: effect on the albino rat, *Ind. Med.,* 9, 296, 1940.

139. **Deichmann, W. and Schafer, L. J.,** Phenol studies. I. Review of the literature. II. Quantitative spectrophotometric estimation of free and conjugated phenol in tissue and fluids III. Phenol content of normal human tissues and fluids, *Am. J. Clin. Pathol.,* 12, 129, 1942.

140. **Sprague, J. B.,** Avoidance reactions of salmonid fish to representative pollutants, *Water Res.,* 2, 23, 1968.

141. EPA, Initial Scientific and Minieconomic Review, No. 24. Methoxychlor, *J. Agric. Food Chem.,* 11, 70, 1976.

142. **Mukherjee, S, and Bhattachaya, S.,** Variations in the hepatopancreatic amalyse activity in fishes exposed to some industrial pollutions, *Water Res.,* 11, 71, 1977.

143. **Halsbrand, E. and Halsbrand, I.,** Veranderungen des Blutbildes von Fischen Infolge toxischer Schaden, *Arch. Fisch Wiss.,* 14, 68, 1963.

144. **Mitrovic, V. V., Brown, V. M., Shurben, D. G., and Berryman, M. H.,** Some pathological effects of sub-acute and acute poisoning of rainbow trout by phenol in hard water, *Water Res.,* 2, 249, 1968.

145. **Davis, H. C. and Hidu, H.,** Effects of pesticides on embryonic development of clams and oysters and on survival and growth of larvae, *Fish Bull.,* 67, 393, 1969.

146. **Jones, J. R. E.,** The reactions of the minnow, *Phoxinus phoxinus* (L.), to solutions of phenol, orthophenol, and para-cresol, *J. Exp. Biol.,* 28, 261, 1951.

147. EPA, Economic Analysis of Pretreatment Standards for Existing Sources of the Electroplating Point Source Category, EPA 440/2-79-031, Office of Analysis and Education, Environmental Protection Agency, Washington, D.C., 1979.

148. **Marce, R. E.,** Cadmium plating still a MUST, in *Industrial Finishing,* Zinc Institute, New York, April, 1978.

149. Industry and Trade Administration, U.S. Industrial Outlook, 1980, U.S. Department of Commerce, Washington, D.C., 1980.

150. **Nass, L. I.,** Stabilization, in *Encyclopedia of Polymer Science and Technology,* Vol. 12, John Wiley & Sons, New York, 1970, 725.

151. Bureau of Mines, Cadmium, Commodity Profiles, Bureau of Mines, Washington, D.C., August, 1979.

152. EPA, Cadmium Additions to Agricultural Levels via Commercial Phosphate Fertilizers, Report SW-718, Environmental Protection Agency, Washington, D.C., September, 1978.

153. Bureau of Mines, Cadmium Quarterly, Mineral Industry Surveys, Bureau of Mines, Washington, D.C., March 4, 1980.

154. Bureau of Mines, Cadmium in 1979, Mineral Industry Surveys, Bureau of Mines, Washington, D.C., January 2, 1980.

155. Mannsville Chemical Products Reporting Service, Chemical Products Synopsis — Phenol, Mannsville, New York, September 1979, 13661.

156. Council on Environmental Quality, Department of Commerce and EPA, The Economic Impact of Pollution Control: A Summary of Recent Studies, U.S. Government Printing Office, Washington, D.C., March 1972, 321.

157. EPA, Development Document for Effluent Limitations Guidelines and Standards for the Iron and Steel Manufacturing Source Category, Draft, Vol. II, EPA 440/1-79/024a, Environmental Protection Agency, Washington, D.C., June, 1979.

158. Office of Management and Budget, Circular A-94, June 22, 1979.

159. EPA, The Economics of Clean Water, Vol. 1, U.S. Government Printing Office, Washington, D.C., 1972, 63.

160. Environmental Protection Agency, Water quality criteria recommendations, *Fed. Regist.,* 45(23), November 28, 1980.

161. Staff communication with the National Kidney Foundation, New York, 1980.

162. Staff communication with the National Cancer Institute, New York, 1980.

163. Bureau of Mines, Cadmium Quarterly, Mineral Industry Surveys, U.S. Bureau of Mines, Washington, D.C., June 1980.

164. **Chase, M. and Malim, T.,** Cadmium profile, *Metal Statistics, 1980,* Fairchild Publications, New York, 1980.

165. **Rade, D.,** Cadmium Pigments for Coloring of Plastics — Still a Must, Paper presented at 2nd Int. Cadmium Conf., Cannes, France, February 6—8, 1978.

166. Federal Reserve Bulletin, Vol. 66, No. 12, December 1980.

167. **Durfor, C. N.,** Public Water Supplies of the 100 Largest Cities in the United States, 1962, Geological Survey Water Supply Paper 1812, U.S. Government Printing Office, Washington, D.C., 1965.

168. Department of Energy, Water Related Planning and Design at Energy Firms, DOE/EV/10180-1, Department of Energy, Washington, D.C., 1980.

169. Environmental Protection Agency, Fuel Cycle Standards, 40 CFR 140, Environmental Protection Agency, Washington, D.C., 1976.

170. *Fed. Regist.,* 44(247), 75928, 1979.

171. Department of Energy, Water Quality Issues and Energy Alternatives, DOE/EV/10154-01, Department of Energy, January, 1981.

INDEX